Systems Biology

Topical Conference at the 2012 AIChE Annual Meeting

Pittsburgh, Pennsylvania, USA
28 October - 2 November 2012

ISBN: 978-1-62276-747-2

Printed from e-media with permission by:

Curran Associates, Inc.
57 Morehouse Lane
Red Hook, NY 12571

Some format issues inherent in the e-media version may also appear in this print version.

Printed by Curran Associates, Inc. (2013)

AIChE
3 Park Avenue
New York, NY 10016-5991

Phone: (203) 702-7660
Fax: (203) 775-5177

www.aiche.org

Additional copies of this publication are available from:

Curran Associates, Inc.
57 Morehouse Lane
Red Hook, NY 12571 USA
Phone: 845-758-0400
Fax: 845-758-2634
Email: curran@proceedings.com
Web: www.proceedings.com

TABLE OF CONTENTS

(42a) Robustness in Nature: Challenges and Opportunities for the Systems Biology Community

Francis J. Doyle III, Chemical Engineering, UCSB, Santa Barbara, CA

Robustness, the ability to maintain performance in the face of perturbations and uncertainty, is a key property of living systems. While 'homeostasis' has long been recognized as an important phenomenon, the molecular and cellular bases of robustness have only recently begun to be understood. Biology and engineering employ a common set of basic 'control' mechanisms to achieve such robust regulation, namely redundancy, feedback control, modularity and hierarchies to ensure robust performance. New systems theoretical approaches to complex engineered systems are required that allow the reverse engineering of general design principles that can provide insights into cellular robustness. While preliminary results are available for simple (low-dimensional, deterministic) biological systems, general tools for analyzing these tradeoffs are the subject of active research. In this talk, I will outline methods that are drawn from the field of control and dynamic systems to generate insights into the functioning of these robust biophysical networks. Examples will be used to motivate problems and methodologies, including two from the medical field (Alzheimer's Disease, Post-traumatic Stress Disorder) and one from Ecology (Synchronization in Annual Spawning of Coral).

(42b) Dynamic Modeling of Metabolism. the Cybernetic Approach

Doraiswami Ramkrishna, School of Chemical Engineering, Purdue University, West Lafayette, IN

Dynamic modeling of metabolism (Song and Ramkrishna[1] [2]) based on the cybernetic concept (Kompala et al[3]; Young and Ramkrishna[4]) is reviewed demonstrating its predictive capabilities. The talk will: (i) introduce its basic tenets, (ii) expound its relationship to other approaches to modeling metabolism, and (iii) to reveal its unique dynamic capabilities. The framework is shown to provide eloquent support for goal-oriented explanations of regulatory biological phenomena within the cause-and-effect methodology of science through experimental conformation of exacting predictions. It is further shown that relatively sparse dynamic data on wild-type strains suffice to predict the dynamic behavior of numerous gene knockout mutant strains providing a promising framework for metabolic engineering with the focus on ***productivity*** of targeted products.

---

[1] Song, Hyun-Seob and D. Ramkrishna, "Prediction of Metabolic Function from Limited Data: Lumped Hybrid Cybernetic Modeling (L-HCM)," *Biotechnol. & Bioeng.,* **106**, 271-284, 2010.

[2] Song, Hyun-Seob and D. Ramkrishna, "Cybernetic Models based on Lumped Elementary Modes Accurately Predict Metabolic Function of Multiple Strains,", *Biotechnol. & Bioeng.,* 127-140, 2011.

[3] Kompala, D.S., D. Ramkrishna., N.B. Jansen and G.T. Tsao, "Investigation of Bacterial Growth on Mixed Substrates. Experimental Evaluation of Cybernetic Models," *Biotech. & Bioeng.,* **28**, 1044-1056, 1986.

[4] Young, J. D. and D. Ramkrishna, "On the Matching & Proportional Laws of Cybernetic Models," *Biotech.Progress,* **23,** 83-99, 2007.

(100a) A Systems-Level Analysis Approach for Identifying Genetic Targets to Treat Biofilm-Forming Pathogens: An Application to Pseudomonas Aeruginosa

Zhaobin Xu, Department of Chemical Engineering , Villanova University, Villanova, PA, Xin Fang, Department of Systems Biology, Henry M. Jackson Foundation, Bethesda, MD, Thomas K. Wood, Department of Chemical Engineering & Biochemistry and Molecular Biology, Pennsylvania State University, University Park, PA and Zuyi (Jacky) Huang, Department of Chemical Engineering & The Center for Nonlinear Dynamics and Control, Villanova University, Villanova, PA

The investigation of biofilm formation is an urgent research theme in systems biology, as biofilms are heavily involved in bacterial chronic inflammatory and infectious diseases (Sauer et al., 2007). It is reported that nearly 65% of all nosocomial infections in the USA are associated with biofilms. Prevention of the initiation of biofilm formation is the most important step for combating biofilm-forming pathogens, as the ability of pathogens to resist antibiotics is enhanced 10 to 1000 times once they form biofilms (Thien and O'toole, 2001). Both experimental approaches (Gerdes et al., 2003) and model-based approaches (Edwards and Palsson, 2000) have been developed to identify genes essential for bacterial growth in the planktonic state. However, inhibition of some essential planktonic-growth genes might prevent the growth of microorganisms in the planktonic state but may not prevent the formation of biofilms. In these cases, microorganisms can still switch from planktonic growth to the biofilm lifestyle before antibiotics completely inhibit biological functions associated with these genes. For example, Zhang et al., 2007 has reported that biofilm formation is a general strategy utilized by pathogens to survive stress such as that imposed by antibiotic treatment. On the other hand, a large amount of genes are involved in biofilm formation. Even though we inhibit some of these biofilm-associated genes, we cannot eliminate pathogenicity as these genes might not be essential for microorganism metabolism. Therefore, the ideal gene targets to treat biofilm-forming pathogens should be those that prevent the formation of biofilms and also eliminate pathogens in the planktonic state.

In this work, we develop a systems-level analysis approach to determine target genes that are essential for bacterial metabolism but which do not induce biofilm formation when they are inactivated. The proposed approach is based on flux balance analysis (FBA) (Edwards and Palsson, 2000), artificial center hint and run sampling (ACHR) (Becker et al., 2007), and hierarchical clustering. Specifically, fluxes of reactions positively associated with biofilm formation are used as sensors to monitor the trend of gene-inhibition mutants to form biofilms. Mechanisms for inhibition mutants of essential genes to form biofilms are investigated by identifying those biofilm-associated reactions whose fluxes surge for most gene-inhibition mutants. Target genes to treat biofilm-forming pathogens are then defined as those essential planktonic-growth genes whose mutants cannot induce the formation of biofilms.

The proposed approach was applied to identify genetic targets to treat Pseudomonas aeruginosa infections, as it is one of the leading causes of nosocomial infections in hospitalized patients and P. aeruginosa displays resistance to a wide array of antibiotics (Cao et al., 2004). The model presented in Oberhardt et al., 2008, was used to predict the trend for P. aeruginosa to switch from the planktonic state to the biofilm growth mode when essential planktonic-growth genes are inhibited. It is interesting to find that that inhibition of most essential planktonic-growth genes induces the formation of biofilms. Specifically, P. aeruginosa tends to survive the essential-gene inhibition treatment (i.e., a pseudo antibiotic-treatment) by mainly increasing the fluxes through 10 reactions that regulate amino acid metabolism, acetate metabolism, hydrogen peroxide, and ABC phosphate transporter. This phenomenon indicates the adaptability of P. aeruginosa toward unfavorable environments. This work presents the first computational approach to address the mechanism under this phenomenon. In addition, three essential planktonic-growth genes, i.e., PA1756 (cysH), PA3686 (adk), and PA2023 (galU), cannot induce biofilm formation when they are inhibited. They constitute gene targets to treat P. aeruginosa as inhibition of them can eliminate pathogens without inducing biofilm formation. All these three genes have been validated by existing data as potential drug targets for their crucial role in the survival or virulence of P. aeruginosa (Bonofiglio et al., 2005; Markaryan et al., 2001; Ren et al., 2005; Williams et al., 2002).

Becker SA, Feist AM, Mo ML, Hannum G, Palsson BO, Herrgard MJ. Quantitative prediction of cellular metabolism with constraint-based models: the COBRA Toolbox. Nat Protoc 2:727-738, 2007
Bonofiglio L, García E, Mollerach M. Biochemical characterization of the pneumococcal glucose 1-phosphate uridylyltransferase (GalU) essential for capsule biosynthesis. Curr Microbiol 51(4):217-221, 2005
Cao B, Wang H, Sun H, Zhu Y, Che M. Risk factors and clinical outcomes of nosocomial multi-drug resistant Pseudomonas aeruginosa infections. J Hosp Infect 57(2): 112-118, 2004
Edwards JS, Palsson BO. Metabolic flux balance analysis and the in silico analysis of Escherichia coli K-12 gene deletions. BMC Bioinformatics 1:1-1, 2000
Gerdes SY, Scholle MD, Campbell JW, et al. Experimental determination and system level analysis of essential genes in Escherichia coli MG1655. J Bacteriol 185(19): 5673-5684, 2003
Markaryan A, Zaborina O, Punj V, Chakrabarty AM. Adenylate kinase as a virulence factor of Pseudomonas aeruginosa. J Bacteriol 183(11):3345-3352, 2001
Oberhardt MA, Puchalka J, Fryer KE, Martins dos Santos VAP, Papin JA. Genome-scale metabolic network analysis of the opportunistic pathogen Pseudomonas aeruginosa PAO1. J Bacteriol 190(8): 2790–2803, 2008
Ren D, Zuo R, Gonzalez Barrios AF, Bedzyk LA, Eldridge GR, Pasmore ME, Wood TK. Differential Gene Expression for Investigation of Escherichia coli Biofilm Inhibition by Plant Extract Ursolic Acid. Appl Environ Microb 71(7): 4022–4034, 2005
Sauer K, Rickard AH, Davies DG. Biofilms and biocomplexity. Microbe 2:347–353, 2007
Thien FCM, O'Toole GA. Mechanisms of biofilm resistance to antimicrobial agents. Trends Microbiol 9(1): 34-39, 2001
Williams SJ, Senaratne RH, Mougous JD, Riley LW, Bertozzi CR. 5'-Adenosinephosphosulfate lies at a metabolic branch point in mycobacteria. J Biol Chem 277(36):32606-32615, 2002
Zhang X, Garcia-Contreras R, Wood TK. YcfR (BhsA) Influences Escherichia coli Biofilm Formation through Stress Response and Surface Hydrophobicity. J Bacteriol 189(8): 3051-3062, 2007

(100b) Engineering ROS Metabolism with Futile Cycles

Kristin Adolfsen, Princeton University, Princeton, NJ and Mark P. Brynildsen, Chemical and Biological Engineering, Princenton University, Princeton, NJ

Reactive oxygen species (ROS) are toxic molecules utilized by the innate immune system and antibiotics to kill bacteria. In addition to their role in host defense, a basal level of ROS is produced inadvertently in all aerobically growing organisms when reduced electron carriers collide with O_2 and transfer their electrons. For this reason, many bacteria have evolved defense capabilities against these toxic species, such as producing detoxification enzymes and antioxidant metabolites. When the level of ROS within a cell exceeds its capacity for detoxification, bacterial death occurs. In previous work, we have shown that increased ROS production potentiates the effects of oxidative stress. An important insight from that study was that ROS production increases as inefficient use or production of ATP increases. In this study, we sought to increase ATP inefficiency by engaging futile cycles and determine if such a perturbation increases sensitivity toward oxidative stress. Futile cycles are sets of reactions whose net activity results in ATP hydrolysis. Using mixed-integer linear optimization, we have identified all futile cycles in *E. coli* metabolism containing ≤ 3 reactions, as well as those that require the expression of only one enzyme in aerobic growth in minimal glucose media. We have experimentally validated the presence of futile cycling and measured an increase in sensitivity to H_2O_2 for strains that are actively futile cycling when compared to strains carrying an inactive form of the cycle-enabling enzyme. The knowledge gained from this project could lead to the development of new therapies to sensitize microbes to immunity and antibiotics.

(100c) Dynamic Modelling As a Tool for Increasing Single-Chain Antibody Fragment Specific Productivity in Pichia Pastoris

Kate Royle, Life Sciences, Imperial College London, London, United Kingdom, David Leak, Biology & Biochemistry , University of Bath, Bath, United Kingdom and Cleo Kontoravdi, Chemical Engineering and Chemical Technology, Imperial College London, London, United Kingdom

Pichia pastoris is a commonly used expression host for heterologous protein production, predominantly because it is amenable to genetic manipulation and can grow to high cell densities in cheap culture media. While considerable yields can be achieved in this way, the specific productivity is relatively low. Consequently, the full impact of this host on industrial biotechnology has not yet been realised.

Previous studies to increase productivity have focused on bioengineering, such as targeting gene and strain characteristics, and fermentation conditions. Whereas the latter has been subject to multiple systematic multivariate optimisations with computational modelling techniques, the former has not. The majority of bioengineering studies have targeted one factor in isolation, and despite comparable strategies have variable outcomes. Here, a dynamic computational modelling approach has been taken to understand how the factors interact and develop a global optimisation strategy.

Initially, a deterministic single-cell model was derived recapitulating the essential features of protein production: transcription and translation were modelled with mass action kinetics, and the folding of nascent protein into its three-dimensional conformation with Michaelis-Menten kinetics based on the foldase Pdi. Additionally an endoplasmic reticulum (ER) stress response, the unfolded protein response (UPR), was included as it is well documented that this is activated during heterologous protein production. In essence, too much unfolded protein in the ER is detected by the receptor Ire1 which triggers a cascade of reactions concluding in an upregulation of proteins to counteract stress. Although this pathway has been modelled in yeast once before, our work has expanded both the breadth and the kinetics. Specifically, stress occurs when the chaperone Kar2 dissociates from Ire1 to bind unfolded protein, allowing excess unfolded protein to bind Ire1. This activates the endonuclease domain of Ire1, which subsequently splices *HAC1* mRNA allowing translation of this transcription factor. The protein-protein interactions have been modelled with equilibrium kinetics, and the splicing of *HAC1* mRNA with Michaelis-Menten kinetics. The transcription factor Hac1 retrotranslocates to the nucleus and the upregulation of *KAR2*, *PDI*, *HAC1* and *E3* has been modelled with Michaelis-Menten kinetics. Finally, the ER associated degradation pathway (ERAD), which acts to remove misfolded protein from the ER and is intertwined with the UPR, has been incorporated - a unique feature making this model more comprehensive than those previously published. Specifically, the binding of Kar2 to misfolded protein has been modelled with equilibrium kinetics and the subsequent degradation with Michaelis-Menten kinetics based on E3, the rate limiting step in the ERAD pathway.

Whilst simulations of the model proved qualitatively accurate, a lack of literature values hindered quantitative accuracy. Consequently, we sought to construct and characterise strains of *P. pastoris* producing a heterologous protein for which single-chain antibody fragments (scFvs) were chosen. Although monoclonal antibodies are currently the dominant biopharmaceutical in manufacture, smaller antibody fragments such as scFvs are expected to feature more heavily in the future. While mammalian cells are most frequently used to produce full antibodies, the smaller fragments are better suited to a microbial host.

Therefore, novel *P. pastoris* strains have been constructed producing two scFvs: BC1, which targets new blood vessels of tumours; and MFE23, which targets gastrointestinal cancers. These genes were cloned into the commercial vector pPICZαA with C-terminal polyhistidine and myc tags, and subsequently integrated into the *AOX1* locus. Expression was driven by the *AOX1* promoter, and secretion conferred by the N-terminal α-Factor from *Saccharomyces cerevisiae*. As there is biological variation in specific productivity, different clones of the same strain can produce varying amounts of scFv. In order to include this variation in the model and help identify relevant factors, both low and high yield clones were isolated. As multiple copies of the heterologous gene can increase productivity, the low and high yield clones were subjected to Southern blot analysis to confirm single copy integrants, allowing the gene copy number in the model to be set to one.

A major source of quantitative inaccuracy in the model was the concentration of two proteins in the ER, the chaperone Kar2 and the foldase Pdi. These proteins were identified as having a large influence over yield; however, literature values varied over three-fold. We therefore sought to accurately quantify these and developed an LC-MS/MS method to do so. Initially, an *in silico* tryptic digestion of the proteins was conducted and the peptides ranked according to their enhanced signature peptide prediction score, a computational method which evaluates their physicochemical properties. MRM methods were written to target the superior peptides, and these were experimentally validated with analysis of a tryptic digest of *Escherichia coli* BL21 DE3 strains overexpressing the proteins. Isotopically labelled versions were subsequently expressed, purified and used to optimise the MRM in terms of collision energy and collision cell exit potential. As the proteins are difficult to express at high yields, however, isotopically labelled peptides were purchased from JPT Peptide Technologies for the experiment.

As the UPR upregulates Kar2 and Pdi, the model required experimental data for the baseline level in unstressed cells, the higher level in stressed cells and the time over which this occurs. As transcription of the *SCFV* genes are specifically induced by changing the carbon source, we have devised an experimental system to generate this profile. Specifically, uninduced strains at an OD_{600} of 1 were incubated at 30°C, 250 rpm for 24 hours. Samples were taken at 22 and 23 hours, at which point the culture is in stationary phase, allowing evaluation of the baseline Kar2 and Pdi values. At 24 hours, scFv expression was induced and further samples taken at 0, 2, 4 and 6 hours. These data provide a complete picture of the unstressed baseline, activation of stress and the final maximum the system can achieve.

Simultaneously, the samples were analysed for the magnitude of the UPR and ERAD pathways. Traditionally, activation of the UPR has been measured using reverse-transcriptase quantitative PCR targeting *KAR2*, *PDI* and also *HAC1* mRNA, and this strategy was implemented here. Indeed, for Kar2 and Pdi it also allowed for interesting comparisons between transcriptional and protein upregulation as it is widely appreciated that they do not always equate. Therefore, primers were designed and optimised to target *KAR2*, *PDI* and *HAC1* in conjunction with those against actin, an internal control for the experiment. Upregulation was calculated with the Pfaffl method, compared to the wild-type GS115. Activation of the ERAD was assessed in a similar way; however, this pathway is poorly described compared to the UPR. Primers were designed to target major contributors, specifically a ubiquitin-activating enzyme (*E1*), a ubiquitin-conjugating enzyme (*UBC7*), two ubiquitin protein ligases (*HRD1* and *DOA10*) and an ATPase (*CDC48*). Interestingly, some studies have suggested that the ubiquitin protein ligases may be substrate specific; consequently analysing both *HRD1* and *DOA10* allowed an insight into whether this is the case for scFvs in *P. pastoris*.

Whilst the above data refined model accuracy, validation was required. Many studies of the UPR chemically induce stress to test hypotheses. This, however, produces pleiotropic effects which are not ideal. Here, we constructed strains which overexpress Kar2 and Pdi both individually and together. The increase in availability of these proteins provides the necessary perturbation to the system, yet they are also expected to have beneficial effects on yield imparting added industrial relevance. Specifically, each gene was cloned into the pIB2 vector under the control of the *GAP* promoter, with an *AOX1* transcription terminator and homology to the *his4* locus for integration. This approach resulted in a constitutive upregulation of the proteins for incorporation into the model. Simultaneous upregulation required a more developed strategy. Ideally, the same promoter was required; however, ongoing research suggests that *P. pastoris* frequently recombines between large regions of homology preventing the use of two copies of the promoter. To circumvent this, both genes were expressed as a single bicistronic mRNA with an IRES sequence joining the two. IRES sequences allow cap-independent translation such that the second gene in the bicistronic mRNA can be translated in addition to the first. Interestingly, there is some evidence that *P. pastoris* may be subject to translational downregulation during ER stress. If this is the case, cap-independent translation will not be hindered. Therefore, *KAR2* was strategically cloned into pIB2 after *PDI* and under control of an IRES sequence from *Saccharomyces cerevisiae*, as it is better able to relieve ER stress. Further RT-QPCR and mass spectrometry analysis confirmed the perturbation and provided a more thorough analysis of the use of IRES sequences in *P. pastoris* – a technique used only once previously.

In conclusion, we have developed a deterministic model which mimics scFv production and the initiation of ER stress pathways. This model has been tested with constructs that over-express Kar2 and Pdi, both validating its predictive capabilities and generating industrially relevant constructs. Finally, it has been simulated to predict the optimal strategy for increasing specific productivity in *P. pastoris*, constituting an integrated modelling and experimental approach to solving a life sciences problem.

(100d) Modeling Intra- and Inter-Kingdom Signaling Through NF-Kb Pathway in Dendritic Cells

Shreya Maiti[1], Robert Alaniz[2], Juergen Hahn[3] and Arul Jayaraman[1], (1)Department of Chemical Engineering, Texas A&M University, College Station, TX, (2)Department of Microbial and Molecular Pathogenesis, Texas A&M Health Science Center, College Station, TX, (3)Biomedical Engineering and Chemical & Biological Engineering, Rensselaer Polytechnic Institute, Troy, NY

The human gastro-intestinal (GI) tract is colonized by ~10^{14} commensal (non-pathogenic) bacteria in close proximity to host epithelial cells [1]. Intra-kingdom (host-host) and inter-kingdom (host-bacteria) communication is integral to proper functioning of GI tract, and helps maintain a delicate balance between a pro-inflammatory response to invading pathogens and tolerance towards commensal bacteria (anti-inflammatory response) [1]. Disruption of this homeostatic balance can lead to indiscriminate phagocytosis of both bacterial and host cells resulting in diseases such as Crohn's disease and Ulcerative colitis [2]. This study focuses on investigating the inter- and intra-kingdom signaling mechanisms involved in maintaining homeostasis in the GI tract through mathematical modeling and experimental validation.

One of the anti-inflammatory metabolites produced by the commensal bacteria is indole. Indole is generated from dietary tryptophan by the commensal bacteria, using the enzyme tryptophanase A, and is released into the lumen of GI tract [3]. Prior work from our lab [3] has shown that microbiota-derived indole attenuates indicators of inflammation by decreasing production of IL-12 and IL-8 while increasing production of IL-10. More importantly, these studies also showed that indole decreased activation of the pro-inflammatory transcription factor NF-kB in intestinal epithelial cells. Since IL-10 is also shown to decrease NF-kB activation [4], these results suggest that NF-kB signaling is a putative mechanism through which inter-kingdom signaling effects are mediated.

In this work, we have developed a mathematical model that represents the signal transduction pathway in dendritic cells through which each host-derived pro-inflammatory signaling and microbiota-driven anti-inflammatory signaling converge at NF-kB. TNF-α is used as the prototypic pro-inflammatory cytokine while IL-10 and indole are used as the anti-inflammatory input to the model. The integrated model has been developed in MATLAB using ordinary differential equations and unknown parameters have been estimated from experimental data. A regulatory feedback loop for TNF-α has been included to account for the effect of *de novo* synthesized TNF-α on NF-kB activation. IL-10 and indole signaling models have been integrated into TNF-α model through common nodes among them. Quantitative experimental data on NF-kB activation has been obtained using dendritic cells transduced with a secreted luciferase reporter system.

The model developed in this study is expected to be helpful in determining the points of intersection of pro- and anti-inflammatory responses and the molecular mediators involved in maintaining homeostasis in the human GI tract, which would in turn help in designing therapeutic strategies for inflammatory disorders.

References:

1. Berg, R.D., *The indigenous gastrointestinal microflora.* Trends in Microbiology, 1996. **4**(11): p. 430-435.

2. Bouma, G. and W. Strober, *The immunological and genetic basis of inflammatory bowel disease.* Nat Rev Immunol, 2003. **3**(7): p. 521-533.

3. Bansal, T., et al., *The bacterial signal indole increases epithelial-cell tight-junction resistance and attenuates indicators of inflammation.* Proceedings of the National Academy of Sciences, 2010. **107**(1): p. 228-233.

4. Dhingra, S., et al., *IL-10 attenuates TNF-α-induced NFκB pathway activation and cardiomyocyte apoptosis.* Cardiovascular Research, 2009. **82**(1): p. 59-66.

(100e) Optimum Perfusion Duration for Machine-Perfused Rat Livers

Sinem Perk[1], Maria-Louisa Izamis[1], Herman Tolboom[2], Basak Uygun[1], Francois Berthiaume[3], Martin L. Yarmush[4] and Korkut Uygun[1], (1)Center for Engineering in Medicine, Massachusetts General Hospital, Harvard Medical School, Shriners Hospital for Children, Boston, MA, (2)Division of Cardiac and Vascular Surgery, University Hospital Zurich, Zurich, Switzerland, (3)Department of Biomedical Engineering, Rutgers, The State University of New Jersey, Piscataway, NJ, (4)Center for Engineering in Medicine, MGH/Harvard Medical School, Boston, MA

With the crisis in donor organ availability, recent studies on liver preservation focus on new, dynamic modalities such as machine perfusion to recover discarded donor organs to a functional state. Compared to static cold storage, machine perfusion provides oxygen and nutrition enabling recovery of the organ. Despite years of research however, little is known about key factors such as how to evaluate the degree of recovery during perfusion and what the ideal perfusion duration is to maximize the graft recovery and transplantation success. Development of such approaches are complicated by the fact that every donor liver is different and therefore it is very likely that the optimum perfusion time, as well as other operating parameters and specific treatments, may be different for every organ.

It is well known that the existing organ preservation systems are limited in duration that the success rate falls dramatically in time. While many studies demonstrate that hours of ex-vivo machine perfusion are necessary for resuscitation of a DCD organ, it is also well known that organ function deteriorates if it is kept in perfusion for too long. Accordingly, the organ viability and transplant success probability likely goes through an optimum, before which the recovery from ischemia is incomplete, and beyond which the organ begins to deteriorate due to damage by the perfusion system.

The aim of this work is the development of an approach that enables assessment of liver viability during perfusion, the prediction of clinical outcome post-transplantation, and the determination of the optimum duration of perfusion.

The ideal method in perfused graft viability assessment would be based on online measurements that can be performed accurately and quickly such as Blood Gas Analyses. The dynamic patterns of organ function during perfusion are evaluated and compared between successful and unsuccessful transplants in a manner that allows a statistical analysis of transplant success, providing an estimate for the likelihood of graft success to the surgeon as a decision support system for transplantation.

With the on-line data obtained during machine perfusion, a suitable methodology for such dynamic organ viability assessment is statistical process monitoring (SPM) based on partial least squares (PLS) methodology and its extension for dynamic processes: multi-way PLS (MPLS). Constructing a statistical model of liver dynamic response to perfusion enables in-silico predictions of the liver trajectory during perfusion. These predictions of the trajectory of the organ in the following hours can then be used to detect any emerging trends that indicate deterioration of organ viability, and therefore enable identifying the optimum perfusion duration, and potentially other operational variables, for each organ based on its function on-line during perfusion.

This work presents a novel approach and a two-step MPLS/PLS algorithm for the prediction of perfused liver survival, based on measurements taken during perfusions of rat livers for 6 and 12 hours, with recipients of the first group surviving and latter group mostly failing. Via an online MPLS analysis, blood gas data taken hourly is used to estimate the post-transplantation survival rates dynamically during perfusion, which are compared with the actual post-transplantation clinical outcomes for each animal. The results demonstrate that with the proposed methodology, graft viability can be monitored and transplant success rate can be predicted online during perfusion with cross-validated sensitivity and specificity over 91%.

(100f) Analyzing the Dynamics of Cell Cycle Transition in Differentiating Embryonic Stem Cells Through an Integrated Experimental and Modeling Approach

Keith Task, Chemical Engineering, University of Pittsburgh, Pittsburgh, PA and Ipsita Banerjee, Chemical and Petroleum Engg, University of Pittsburgh, Pittsburgh, PA

Embryonic stem cells (ESC) have the potential to be used in many therapeutic applications due to their ability to expand in an undifferentiated state while being able to differentiate to any cell type under specific conditions. One important characteristic vital to this behavior is the cell cycle. In somatic cells, the cell cycle is characterized by a lengthy G1 (growth) phase and a long doubling time. ESC are typically characterized by a shortened G1 phase, with an increased proportion of cells residing in the S phase (DNA replication). This leads a shorter time between mitotic events. These traits are crucial for ESC to facilitate quick self-renewal and proliferation. Although these unique cell cycle traits are central to ESC, attempts towards a thorough quantitative analysis of the cell cycle behavior in human ESC (hESC) and in reprogrammed induced pluripotent cells (iPS) is lacking thus far. Furthermore, little work has been done in the area of mechanistic information and dynamics of cell cycle transition from undifferentiated to germ layer. In this project we are using an integrated experimental and computational approach to investigate cell cycle traits with different hESC and iPS cell lines, both at the singular cell and population levels. Furthermore, we investigate the dynamics of cycle behavior transition when differentiation is induced.

Two different hESC lines (the use of which we have approval for) were used for the study, H1 and H9, as well as three iPS lines (Y1 and hFib2-iPS5, derived from dermal fibroblasts, and a type I diabetes specific line). Proportions of the cell cycle were quantified with the DNA stain propidium iodide (PI) and flow cytometry. Further characterization of the cell cycle was performed with Brdu, an S-phase proliferation maker, to extract information regarding the length of cell cycle phases. Quiescent cell state was determined by the lack of Ki67 stain. The undifferentiated cells were first analyzed to determine base level cycle phase characteristics. To study how this heterogeneous population is generated from a homogeneous starting point, the pluripotent cells were synchronized in the G2 phase by Nocodazole. Cells were released from cell synchronization, and the DNA content was analyzed every two hours to determine, temporally, how the cells revert to their asynchronous behavior. In the differentiation studies, endoderm was induced by supplementing culture media with Activin A. Four different induction pathways were compared: Activin A supplemented with either Wnt3a, Bmp4, Fgf2 or PI3K inhibition. PI, Brdu, and Ki67 stains were performed at different time points for each condition to determine the dynamics of the cell cycle phase shift. To determine the extent of differentiation, both quantitative polymerase chain reaction (qPCR) and intracellular flow cytometry were utilized. Furthermore a stochastic population based model was developed to extract mechanistic information from the experimental observations.

The simulated cell cycle data was compared to the experimental observations, and through this comparison probabilistic parameters and characteristics associated with ESC were determined. Only certain probability distributions and moments (e.g. mean and variance) were able to recapitulate experimental dynamics. It is shown that the simulated data can describe the experimental data well, being able to generate the PI flow cytometry histograms, both of the asynchronized pluripotent cells and the dynamics after release from synchronization. The doubling time predicted by the model was validated by time-lapse microscopy and Brdu analysis. Furthermore, by performing the analysis on differentiating cells, it is shown that different induction conditions show different behavior in terms of cell cycle, with different dynamics associated with the transition from "pluripotent cycle" (shorter doubling times and reduced G1 time) to "differentiated cycle." Our stochastic population model is able to draw mechanistic insight regarding both the nature of the cell cycle and of differentiation from experimental observations. Although applied to undifferentiated and germ layer cells, the model could be extended to later stages of differentiation.

(100g) Analysis of EGFR Flux Through Different Endocytic Pathways in Cancer Cells with Elevated EGFR Expression

Alice J. Macdonald, Bioengineering, University of Pennsylvania, Philadelphia, PA and **Matthew J. Lazzara**, Chemical and Biomolecular Engineering, University of Pennsylvania, Philadelphia, PA

In cancer cells characterized by elevated expression of the epidermal growth factor receptor (EGFR), perturbations to the kinetics of ligand-mediated EGFR endocytosis are frequently observed. Such perturbations can have important consequences for the ability of EGFR to activate downstream signaling pathways and may therefore affect cancer cell growth rates and response to therapeutics. Given the existence of endogenous mechanisms which both promote and impede EGFR endocytosis in response to EGFR activation, the question arises as to how these processes act in concert to determine rates of EGFR endocytosis in cancers with elevated EGFR expression. To quantitatively address this question in the context of non-small lung cancer (NSCLC) cells with elevated EGFR expression, we developed a novel computational model to parse the effects of Sprouty2, an antagonist of EGFR endocytosis through the CBL pathway, and MIG6, which promotes non-CBL-mediated EGFR endocytosis, on overall EGFR endocytosis kinetics. Our computational framework consisted of a set of coupled ordinary differential equations describing the kinetics of EGFR ligand binding, EGFR dimerization, and EGFR endocytosis through pathways involving CBL (which is sequestered by Sprouty2), MIG6, or neither protein. To determine the unknown rate constants describing the flux of EGFR through the different endocytosis pathways, we created a panel of H1666 and PC9 NSCLC cell lines with knockdown of Sprouty2, MIG6, or both proteins. Whereas H1666 cells express wild-type EGFR, PC9 cells express a constitutively active EGFR mutant which displays impaired ligand-mediated endocytosis. In addition to measuring rates of EGF-mediated EGFR endocytosis in these cell lines, we also measured EGFR expression levels and the fraction of endocytosed EGFR which was recycled to the cell surface. In both cell lines, SPRY2 knockdown promoted EGFR endocytosis and MIG6 knockdown impeded endocytosis. With SPRY2 knockdown alone or together with MIG6 knockdown, we also observed a marked reduction in EGFR expression in both cell lines. Based on these data, our model results suggested that the effects of Sprouty2 knockdown on EGFR endocytosis were almost completely attributable to changes in EGFR expression, a prediction which we validated by reconstituting EGFR expression in H1666 and PC9 cells with Sprouty2 knockdown. In addition, our model results suggested that the MIG6-mediated internalization pathway plays a surprisingly significant role in setting EGFR endocytosis kinetics of wild-type EGFR in H1666 cells. Not surprisingly, the CBL-mediated pathway was also found to play a significant role in EGFR endocytosis in H1666 cells. By contrast, the model results suggested that the CBL-mediated pathway plays essentially no role in mutant EGFR endocytosis in PC9 cells and that MIG6 may promote mutant EGFR endocytosis less efficiently than wild-type EGFR endocytosis. In addition to these specific conclusions regarding wild-type and mutant EGFR endocytosis in NSCLC cells, the results of this work demonstrate the utility of our computational and experimental approach for identifying the most relevant determinants of EGFR endocytosis in different cellular contexts. This approach may be of general use in identifying optimal ways to regulate EGFR endocytosis kinetics for therapeutic purposes.

(100h) A Stochastic Model of Glycosylation of Monoclonal Antibodies

Devesh Radhakrishnan[1], Andrew Bitner[1], Melissa St. Amand[1], Kevin Tran[1], Anne S. Robinson[1,2] and Babatunde A. Ogunnaike[1], (1)Department of Chemical and Biomolecular Engineering, University of Delaware, Newark, DE, (2)Department of Chemical and Biomolecular Engineering, Tulane University, New Orleans, LA

Most proteins expressed in eukaryotic cells undergo the post-translational modification known as glycosylation in which a glycan (sugar moiety) is attached to the core of the protein. The addition of the sugar can either be to the amine group on the protein (N-glycosylation) or to the hydroxyl group (O-glycosylation). Of these, N-glycosylation, in which the glycan is added to the core of the protein in the lumen of the endoplasmic reticulum of the cell, followed by subsequent modification in the Golgi apparatus, has the greater impact on protein structure and properties. (For instance, the different N-glycans attached to the protein affect immunogenicity, *in vivo* half-life, and the receptor recognition of the protein.) Several commercially relevant biopharmaceuticals such as monoclonal antibodies (mAbs), erythropoietin, etc., are glycosylated proteins whose final product quality characteristics depend on the distribution of the different glycoforms. In order to ensure consistent product quality, therefore, regulatory agencies are encouraging manufacturers to monitor and control the quality of the product during the course of manufacturing. This requires the implementation of an appropriate control strategy, and developing such a strategy for effective control of glycosylation requires a mathematical model that represents the process of glycosylation adequately. Several models exist for predicting the distribution of glycoforms in different cellular systems[1-4] on the basis of overall production of protein in the culture. However, the glycan addition to a protein is a molecular phenomenon occurring within the microscopic volumes of vesicles in the Golgi apparatus, essentially invalidating the continuum assumption for glycan concentrations. Under such circumstances, a stochastic model will provide a more appropriate representation of the mechanisms of glycosylation and glycoprotein formation.

In this work, we develop what is, to our knowledge, the first stochastic model of glycosylation, and discuss the key challenges in the implementation of such a model, along with our approach to handling these challenges. We present results of our stochastic model's prediction of the glycoform concentration profiles and compare them to corresponding deterministic model predictions.

References

1. Umaña, P.; Bailey, J. E. Biotechnology and Bioengineering 1997, 55, 890-908.

2. Krambeck, F. J.; Betenbaugh, M. J. Biotechnology and Bioengineering 2005, 92, 711-728.

3. Hossler, P.; Mulukutla, B. C.; Hu, W. S. Plos One 2007, 2.

4. Jimenez Del Val, I.; Nagy, J. M.; Kontoravdi, C. Biotechnology progress 2011, 27, 1730-1743.

(116a) A Systems Strategy for Engineering Families of Orthogonal RNA Transcription Regulators for Engineering Gene Networks

Melissa Takahashi and **Julius B. Lucks**, School of Chemical and Biomolecular Engineering, Cornell University, Ithaca, NY

A central goal of synthetic biology is to develop a toolkit of building blocks that can be used to engineer whole biomolecular and cellular systems from the bottom up. A key component of this toolkit are biological parts that regulate gene expression in ways that can be used to construct networks that integrate and propagate cellular information according to design. Recently, non-coding RNA-based gene regulatory mechanisms have emerged as powerful and versatile substrates for regulatory parts development [1-3]. One example of an RNA-based regulator is the transcription attenuator engineered from the *S. aureus* plasmid pT181 [4]. This antisense RNA-mediated mechanism has already been shown to be able to evaluate genetic NOR-like logics in front of a single target, and is the only RNA-based mechanism that can directly propagate RNA signals in simple transcriptional networks. However, the construction of more sophisticated RNA-based circuitry requires larger numbers of orthogonally acting attenuators for independent control of multiple targets within the network design. Interestingly, the recent discovery by systems biologists that bacterial genetic networks commonly utilize multiple non-coding RNA regulators in critical functional positions [5] suggests that nature may have already solved the RNA regulator orthogonality problem. We have therefore devised a strategy to expand our library of attenuators by creating chimeric fusions between the pT181 attenuator transcriptional expression platform, and RNA binding domains from other families of natural non-coding RNAs.

The pT181 attenuator is an RNA sequence in the 5' untranslated region of a transcript that regulates transcription elongation through structural changes mediated by antisense-RNA binding. This binding is initiated between a loop structure of the attenuator, and the complementary loop structure of the antisense RNA. Therefore, we hypothesized that new orthogonal chimeric attenuators could be engineered by replacing the RNA-loop sensing region of the pT181 attenuator with RNA-loop sensing regions from a number of naturally occurring antisense RNA-mediated translation regulatory mechanisms. A systematic investigation of fusion position with three different translation systems that utilize loop-loop interactions led to the addition of two orthogonal chimeric attenuators to the existing library. In addition, we created an orthogonal chimeric fusion from a translation mechanism that utilizes a loop-linear RNA-RNA interaction mechanism. Structural analysis of both functioning and non-functioning chimeric attenuators with a high-throughput chemical probing technique called SHAPE-Seq revealed insights into the structural requirements of the chimeric junction. This work is leading to the development of RNA structure/function design rules that will allow the creation of a bioinformatic systems approach to search through natural RNA networks to computationally design larger numbers of orthogonal RNA transcription regulators. This will expand our ability to construct larger synthetic RNA networks, as well as contribute to a systems-level understanding of RNA structure/function modularity that will lead to a deeper understanding of RNA's role in biology.

[1] Buskirk, A. R., Kehayova, P. D., Landrigan, A., Liu, D. R., In vivo evolution of an RNA-based transcriptional activator. *Chem Biol* 2003, *10*, 533-540.

[2] Bayer, T. S., Smolke, C. D., Programmable ligand-controlled riboregulators of eukaryotic gene expression. *Nat Biotechnol* 2005, *23*, 337-343.

[3] Isaacs, F. J., Dwyer, D. J., Ding, C., Pervouchine, D. D., *et al.*, Engineered riboregulators enable post-transcriptional control of gene expression. *Nat Biotechnol* 2004, *22*, 841-847.

[4] Lucks, J. B., Qi, L., Mutalik, V. K., Wang, D., Arkin, A. P., Versatile RNA-sensing transcriptional regulators for engineering genetic networks. *PNAS* 2011, *108*, 8617-8622.

[5] Gottesman, S., Storz, G., Bacterial Small RNA Regulators: Versatile Roles and Rapidly Evolving Variations. *Cold Spring Harb Perspect Biol* 2011, *3*, 1-16.

Cardinale Stefano, PhD
Postdoctoral researcher
Prof. Adam Arkin Laboratory
Lawrence Berkeley National Laboratory – Physical Biosciences Division
Department of Bioengineering - UCB
Berkeley, CA 94720
Email: scardinale@lbl.gov
tel. 510 – 643 - 5683

Systematic analysis of host – genetic circuit interaction reveals host-specific epistatic effects on gene expression in *E. coli*

The genotypic and phenotypic diversity among organisms, or even cells in the same population, often undermines the function of synthetic genetic devices. In addition, advances in genetic engineering, such as recombineering or *in vivo* evolution, enable fast modification of dozens and possibly hundreds of genes at once in bacteria. The efficient manipulation of the host genome can potentially have a strong impact on global gene expression and regulation, and affect the operation of designed genetic circuits.

We are investigating how the genotype of the host, through phenotypical and regulatory factors, affects the output of genetic circuits in the bacterium *E. coli*. Synthetic genetic circuits comprising three bright, fast-folding reporters, expressed from constitutive promoters, are used as model. We first measured reporter output in several parent strains of *E. coli* and found that the output from our model circuit strongly depends on the specific host cell in which it functions. We have measured several cellular physiological factors, such as the rate of growth or the ribosome content, and modeled the relationship between expression level and these variables (Fig. 1).

We continued our investigation with a pilot group of 88 single-gene deletion mutant strains and found that the activity of our model circuits is not invariant to even single-gene differences in the host genetic make-up. We identified various genes that, although have no known function in gene expression regulation, significantly affect circuit activity. In particular, we established a link between nutrient sensing, flagellum regulation and global translational regulation, which directly perturb a genetic device. We are in the process of classifying global and sequence-specific sources of disruption of circuit function. For instance, we identified sequence-specific disruption of our model genetic device mediated by the lag (recovery) phase of growth in the *E. coli* strain HB101 and knockout strains for genes *fis*, *pfkA*, *gnd* and *cysA* (Fig. 2). We have characterized a multivariate regression model that recapitulates ~40% of gene expression variance in the strains utilized in this study. In addition to cellular functions such as translational capacity (cellular ribosome concentration) and cell fitness, we found that a measure of the length of lag phase can significantly increase the prediction of exponential protein concentration in *E. coli*.

We are currently expanding our analysis to include all ~3800 strains of the KEIO collection. This work is elucidating the role of each gene of the *E. coli* genome in cellular homeostasis from the viewpoint of synthetic genetic circuit functionality. Our systematic analysis of host-circuit contextual effects will likely identify novel epistatic interactions in *E. coli*, better understand the genotype-to-phenotype association and improve predictability of forward engineering with synthetic devices in synthetic biology.

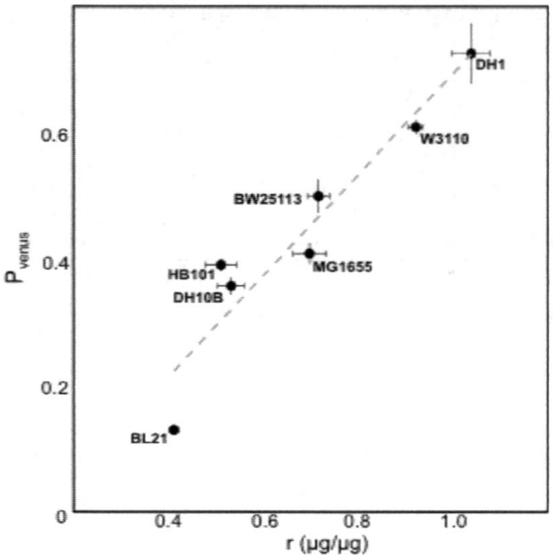

Fig. 1 Linear relationship between translational capacity (r) and protein expression (P) in various *E. coli* parent strains

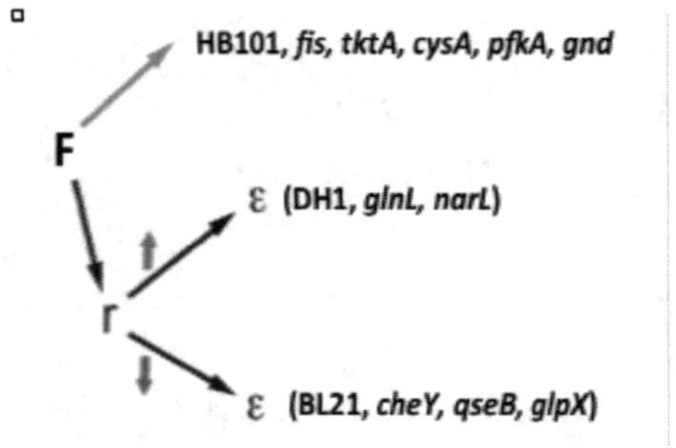

Fig. 2. Cell functions F can affect gene expression ε globally through translational regulation (*r*), or indirectly through undefined mechanisms (gray arrow). We identified the length of lag phase of growth as one of them.

(116c) Molecular Crowding Shaping of Gene Expression Dynamics

Cheemeng Tan[1], Saumya Saurabh[2], Marcel Bruchez[2], Philip R. LeDuc[3] and Russell Schwartz[4], (1)Lane center for computational biology, Carnegie Mellon University, Pittsburgh, PA, (2)Chemistry, Carnegie Mellon University, Pittsburgh, PA, (3)Mechanical Engineering, Carnegie Mellon University, Pittsburgh, PA, (4)Biological Sciences, Carnegie Mellon University, Pittsburgh, PA

Synthetic biology studies have unraveled insights into gene circuit dynamics, but often neglected the impact of non-transcriptional factors, including circuit-host interactions, epigenetic factors, and intracellular microenvironments. An obvious and important non-transcriptional factor is molecular crowding, which refers to the volume exclusion effect resulting from the packing of high-density macromolecules into constrained intracellular spaces. To date, it remains unclear if and how molecular crowding can impact dynamics of gene circuits. We addressed this question by using a multi-scale synthetic biology approach by integrating single-molecule experiments, cell-free expression systems, and artificial cells. Specifically, we found that the size of crowding molecules uniquely affects both the diffusion T7 RNA polymerase and its binding to a T7 RNAP promoter. Based on the single-molecule results, we further showed that the impact of molecular crowding on gene circuits could be enhanced by weak genetic components and maximized by a negative feedback loop. Importantly, molecular crowding confers robustness of gene expression against perturbations of gene environment in both cell free systems and artificial cells. Our results have broad implications on the engineering of synthetic biological systems, gene regulation, and evolution.

(116d) The Single Cell Distribution of Plasmid Copy Numbers

James Boedicker, Applied Physics, Caltech, Pasadena, CA, Franz Weinert, Caltech, Pasadena, CA and Rob Phillips, Applied Physics and Mechanical Engineering, Caltech, Pasadena, CA

Plasmids are a critical piece of many bacterial genomes, and also have an important role in synthetic biology. The ability to manipulate parameters such as gene copy number in a predictable way is desirable for many synthetic gene circuits, and it is often important for biological systems to be robust against variability in gene dosage. The mean number of copies per cell is well known for most plasmids, and some plasmids can be present in hundreds of copies per cell. However less is known about the cell to cell variability of plasmid copy number across a population of cells. The shape of the single cell distribution of plasmid copy numbers can have important implications for gene expression and will help us gain insight into the underlying mechanisms that regulate plasmid copy number, namely plasmid replication and partitioning.

In order to more quantitatively understand the processes which influence plasmid copy number, we measured the single cell distribution of plasmid copy numbers for a population of *E. coli* harboring a high copy number plasmid containing a ColE1-like origin. We have developed a method to analyze the plasmid content of single cells which combines FACS single cell sorting with digital PCR to count the number of plasmids in each cell.

Our results indicate that the distribution of this plasmid at the single cell level is much wider than would be expected from a simple model of binomial partitioning of the plasmids at division and is skewed towards low copy numbers. Using our measured distribution as a guide, we develop and test theoretical models of plasmid replication and partitioning to explore the range of models which would result in a similar distribution of single cell plasmid counts. We also implement dynamic and equilibrium models to examine how the plasmid distribution influences gene expression. From these models, we predict the consequences of different plasmid regulatory mechanisms, such as plasmid loss rates and gene expression patterns, and test these predictions using single cell microscopy. We demonstrate how precise measurements of single cell plasmid distributions enabled us to dissect the mechanisms and consequences of plasmid regulation.

(116e) Exploring Bacterial Sugar Consumption At the Single-Cell Level Using An Integrated Computational and Experimental Approach

Konstantinos Biliouris, Chemical Engineering and Materials Science, University of Minnesota, Minneapolis, MN, Taliman Afroz, Chemical and Biomolecular Engineering, North Carolina State University, Raleigh, NC, Chase L. Beisel, Department of Chemical and Biomolecular Engineering, North Carolina State University, Raleigh, NC and Yiannis Kaznessis, Department of Chemical Engineering and Materials Science, University of Minnesota, Minneapolis, MN

L-arabinose is a five-carbon monosaccharide, found in hemicelluloses (biomass) and abundant in nature. It has recently received increasing interest as hemicelluloses constitute the raw material for production of commodity chemicals and fuels using microbes and must be broken down in this production process. The most common microorganism used for breaking down L-arabinose and producing fuels and chemicals is *E. coli.*

*E. coli*possesses the arabinose system which consists of different operons that are responsible for the uptake and degradation of arabinose. The main operon, called L-arabinose operon, comprises of a regulatory protein, which controls gene expression in an arabinose-dependent fashion, as well as genes that encode proteins involved in the catabolism of arabinose. It has been demonstrated that when the genes important for arabinose degradation are deleted, administration of arabinose results in an all-or-nothing behavior. In particular, the cell population is divided in a subpopulation with saturated arabinose concentrations and a subpopulation with no arabinose at all.

Even though the arabinose consumption mechanism in *E. coli*has been thoroughly investigated, most of the studies have explored its dynamics in the absence of arabinose consumption, i.e. deleting the genes associated with arabinose degradation. We are using a combination of computational and experimental work to interrogate the dynamic behavior of the arabinose system both in the presence and absence of arabinose consumption proteins. We are developing a detailed stochastic mathematical model accounting for the biomolecular interactions underlying the arabinose system. We monitor the evolution of cell behavior over time using fluorescence microscopy and directly compare the experimental phenotype with the simulation results. The simulations are performed using an algorithm which is a fusion of a stochastic-discrete and a stochastic-continuous approach.

Importantly, we are observing an intriguing dynamic behavior which differs significantly from the well characterized all-or-nothing behavior. Overall, our study provides new insights into arabinose consumption at the single cell level. Therefore, our findings may contribute to reverse-engineering sugar consumption in *E. coli* or designing complex synthetic biological circuits whose functionality is analogous to arabinose system.

(116f) Investigating Heterogeneous System Performance of Synthetic Myosins Computationally

Paul Egan[1], Jonathan Cagan[1], Christian Schunn[2] and Philip R. LeDuc[1], (1)Mechanical Engineering, Carnegie Mellon University, Pittsburgh, PA, (2)Psychology, University of Pittsburgh, Pittsburgh, PA

Myosin motor proteins are ubiquitous in natural biological systems and convert chemical energy into mechanical force. Myosins often operate in large groups, with their collective interactions contributing to an emergent system performance that varies according to the isoforms present. These systems are possible to implement in novel myosin-based technologies and the directed design of synthetic myosins could improve system performance. Empirical studies have demonstrated that myosin behaviors are tunable via alterations of specific molecular structures, but predicting the effects of alterations to individual myosins on system performance is difficult due to the large number of influential variables. Therefore, we developed a computational method for procedurally designing individual myosins and systems, thus characterizing the influence of each parameter in the design space. Our findings suggest that the considered design parameters affect system performance differently and often in non-intuitive directions. Additionally, we find that heterogeneous systems (i.e. multiple myosin isoforms working together) have unique and potentially superior performance when compared to homogeneous systems (i.e. single myosin isoforms). The developed approach offers a novel perspective on the complexities of myosin systems and the techniques are extendible for investigating many other biological systems at the interface of chemistry and mechanics.

Our computational model enables the design of myosin systems from the bottom-up, and predicts the emergent system response to force perturbations. This approach allows for separate investigation of variables at the levels of individual myosins and the system. We consider four inputs that represent mechanical or chemical behaviors when configuring myosins and consider discrete numbers of each isoform as the system input. The approach also enables the modeling of heterogeneous systems, which exist in nature but have had little empirical or computational investigation.

We derive a mechano-chemical model based on the swinging lever arm theory for mapping structures manipulated through synthetic myosin experiments to specific behaviors. The swinging lever arm theory proposes that a myosin chemically attaches to a filament, then mechanically exerts positive force as a linearly elastic element during its power-stroke followed by negative force before detaching chemically. We consider the configuration of four structures as inputs: (1) the length of the lever arm and (2) the rotation of the converter domain, which influence the mechanical behaviors of a myosin during its power-stroke and the configuration of the (3) loop one and (4) loop two substructures of the myosin head domain, which influence a myosin's chemical behaviors.

We use differential equations to determine the behavior of each myosin averaged over their entire cycle and then sum the contributions of all myosins to determine the system performance. The empirically measured force-velocity response of a known myosin isoform is used to validate our model, and considered as a datum myosin design. We then extrapolate novel myosin designs and compare their performance to the datum. It is not possible to validate all extrapolated designs, but for the empirical evidence available, our model produces a good fit for maximum system velocity when the lever arm length and detachment rates are extrapolated.

The design of myosins and their numbers in a system are procedurally varied with respect to one variable at a time, which allows for isolating how each parameter affects system performance. The maximum force of a system, maximum filament velocity, system stability, energy utilized, and volume of a system are all investigated and it is found that no two variables affect the system in the same manner. We consider systems that have at least two myosins attached on average stable, therefore the system possess a low probability of system dissociation due to diffusion and a high likelihood of consistent filament trajectory. Robustness of these systems with respect to function (ability to operate under a number of different conditions) is also considered, and it is found that more robust systems generally require more energy use by the system and more volume.

When heterogeneous systems are investigated, it is found that the myosins interacting within the system produce nonlinear performance curves that are not possible to recreate with any homogenous system. External forces are then applied to systems and the response with respect to each performance parameter is determined numerically. It is found that when the performance of a system is considered over a range of conditions, that the unique characteristics of heterogeneous systems have increased performance over any possible homogenous system for certain applications. For instance, one heterogeneous system had decreased energy usage at lower velocities (i.e. greater performance), yet allowed for a higher maximum system velocity with consideration to stability (i.e. greater robustness) in comparison to the optimal homogenous system under the same conditions.

These results provide novel insights for why nature may favor molecularly heterogeneous systems for functions such as muscular contraction. An often stated benefit of myosin heterogeneity in muscles is the ability for muscle performance to adapt over time to external stimuli by tweaking myosin concentrations. However, our findings suggest that the heterogeneity may also provide increased performance for conditions at a specific point in time. Proving that heterogeneity provides increased performance is difficult through empirical studies, because it is difficult to decouple the influences of each myosin from the system as a whole. Thus, our computational approach has provided novel insights for how the configuration of individual myosins affects system performance. Our model has great potential to aid the design of myosin-based technologies, by enabling designers to predict how alterations of individual components as well as the number of each component present affect the performance of a system as a whole. These techniques for decoupling individual molecule behaviors from system performance are crucial for aiding engineers to find the optimal design of the system, and are extendible for investigating heterogeneous biological systems in general, such as the many other classes of motor proteins.

(116g) Construction of Genetic Programs by Layering Logic Gates in Single Cells

Tae Seok Moon[1,2] and Christopher A. Voigt[1], (1)Synthetic Biology Center, Department of Biological Engineering, Massachusetts Institute of Technology, Cambridge, MA, (2)Department of Energy, Environmental, and Chemical Engineering, Washington University in St. Louis, St. Louis, MO

Many applications require cells to integrate multiple environmental signals and to implement synthetic control over biological processes. Genetic circuits enable cells to perform computational operations, interfacing biosensors and actuators. Despite advances in the rational construction of genetic circuits, synthetic biologists are still limited by the lack of well-characterized orthogonal parts and predictive computational models. Here, we harness a toolbox of genetic parts from bacteria to construct three 2-input AND gates. To this end, genetic parts have been recruited from Type III Secretion System (T3SS) where activator-chaperone pairs interact to regulate virulence gene expression. Such pairs have been assembled in *E. coli* and their expression has been tuned by saturation mutagenesis of promoters and RBSs. These engineered 2-input AND gates have been fully characterized, modified (via directed evolution) to make them orthogonal, and rationally layered to make complex genetic programs, the largest of which is a 4-input AND gate that consists of 11 regulatory proteins. In addition, transfer function models have been developed to computationally analyze the AND gates. This work represents a step toward constructing programmed cells that are able to process multiple input signals and to produce desirable outputs for real-world applications.

(175b) Elucidating the Genetic Architecture of Isobutanol Tolerance in Escherichia Coli Through Targeted Genome Engineering and High Throughput Screening

Jeremy J. Minty[1], Jihyang Park[1], Lawrence Lai[1], Ted A. Zaroff III[1], Brian N. Johnson[1], Mark A. Burns[1], Harris Wang[2], George M. Church[3] and Xiaoxia Lin[1], (1)Department of Chemical Engineering, University of Michigan, Ann Arbor, MI, (2)Wyss Institute for Biologically Inspired Engineering, Harvard University, Boston, MI, (3)Genetics, Harvard Medical School, Boston, MA

Understanding the genetic architecture of complex phenotypes is of great fundamental interest and has important ramifications in biotechnology. Metabolic engineering efforts have enabled microbial production of higher molecular weight alcohols as next-generation biofuels, but toxicity limits production. Microbial stress tolerance is a complex multigenic trait intractable to traditional genetic study and rational engineering efforts. Most approaches to improving stress tolerance are therefore combinatorial, following a strategy of generating genotypic and phenotypic diversity in a population, then characterizing isolates with the desired properties. However, present methods explore relatively small genotype spaces and often fail to capture epistatic interactions between distal genetic loci. Guided by evolutionary-genomic studies, we are using Multiplex Automated Genome Engineering (MAGE) to generate combinatorial libraries of 38 mutations associated with isobutanol tolerance in *E. coli*. Using a high-throughput microfluidic screening platform, we characterize phenotype distributions within the libraries and isolate variants with improved isobutanol tolerance. Isolates are further characterized via detailed phenotyping and genotype analysis, allowing for systematic mapping of isobutanol tolerance phenotypes and genotypes. Our work reveals prevalent epistasis between genetic loci and provides insights into biochemical mechanisms of tolerance, as well as generating improved strains of *E. coli* that may be immediately useful for isobutanol production.

(175c) Towards a Semi-Synthetic Stress Response System to Engineer Solvent Tolerance

Kyle Zingaro, Chemical Engineering, University of Delaware, Newark, DE and Eleftherios T. Papoutsakis, Dept. of Chemical Engineering, Delaware Biotechnology Institute, Newark, DE

Many desirable cellular phenotypes are affected by multiple genes and systems functioning together. These so-called complex phenotypes require new approaches which utilize multiple genes simultaneously to engineer an optimal result. To that end we have utilized a method employing multiple coexisting plasmids to generate a semi-synthetic stress response system to engineer an *Escherichia coli (E. coli)* strain capable of tolerating high levels of toxic solvents. We found that not only can overexpression of multiple genes produce a beneficial phenotype for the complex trait of alcohol tolerance but the expression level of these genes, regulated through induction level of inducible promoters and plasmid copy number, can have a significant effect on the overall success of a tolerance system. Through plasmid based expression of autologous heat shock proteins in *E. coli* we have produced a tunable stress response system to improve tolerance to a variety of toxic chemicals.

Overexpression of the autologous GroESL proteins in *E. coli* enhanced cell growth to all alcohols tested, including a 12-fold increase in total growth in 48-hour cultures under 4% (v/v) ethanol, a 2.8-fold increase under 0.75% (v/v) n-butanol, a 3-fold increase under 1.25% (v/v) 2-butanol, and a 4-fold increase under 20% (v/v) 1,2,4-butanetriol. GroESL overexpression resulted in a 9-fold increase in CFU numbers compared to a plasmid control strain after 24 hours of culture under 6% (v/v) ethanol, and a 3.5-fold and 9-fold increase for culture under 1% (v/v) n-butanol and i-butanol, respectively. Simultaneous overexpression of GrpE and GroESL produced a 2-fold increase in viable cells from a double plasmid control in 5% (v/v) ethanol, overexpression of GroESL and ClpB on multiple plasmids produced a 11.3-fold increase in viable cells after 24 hours of exposure to 5% (v/v) ethanol, and overexpression of GrpE, GroESL, and ClpB on a single plasmid produced a 2.5-fold increase in viable cells after 24 hours of exposure to 7% (v/v) ethanol. Through the production of engineered strains designed to increase expression of the heat shock proteins GroES, GroEL, ClpB, and GrpE we have produced a semi-synthetic stress response system capable of greatly increasing the tolerance of *E. coli* to a selection of useful alcohols, and potentially providing a platform to generate tolerance to a wide variety of toxic chemicals.

(175d) Improvement of D-Glucaric Acid Production From a Synthetic Pathway in Escherichia Coli

Eric Shiue and Kristala Jones Prather, Chemical Engineering, Massachusetts Institute of Technology, Cambridge, MA

Improvement of D-Glucaric Acid Production from a Synthetic Pathway in *Escherichia coli*

Eric Shiue, Kristala Jones Prather

D-glucaric acid is a naturally occurring compound that has been investigated for a number of potential uses, including cancer treatment, cholesterol reduction, and production of biodegradable polymers. Additionally, D-glucaric acid has been labeled a "top value-added chemical from biomass" by the US Department of Energy. A pathway for the biosynthesis of D-glucaric acid from glucose has been constructed in *E. coli*, consisting of three heterologously expressed natural enzymes: *myo*-inositol-1-phosphate synthase (INO1) from *Saccharomyces cerevisiae*, *myo*-inositol oxygenase (MIOX) from mouse, and uronate dehydrogenase (Udh) from *Pseudomonas syringae*. Using this pathway, glucaric acid titers of 1.13 g/L were achieved following optimization of culture and induction conditions.[1]

Characterization of the pathway's constituent enzymes has revealed MIOX to be the rate-limiting enzyme. Achievement of industrially relevant titers of D-glucaric acid therefore hinges on improving the activity of MIOX. Various strategies have been explored to improve MIOX activity, including timed induction of pathway enzymes, directed evolution of MIOX for improved activity, and MIOX-MBP fusions for enhancement of soluble expression. We will present experimental results on the application of these strategies towards enhancing MIOX activity and glucaric acid productivity.

---

[1] Moon, TS et al. *Appl. and Env. Microbiology* 75:589-595

(175e) Integrating Systems and Synthetic Biology for Engineering Chemical Production in Bacteria

Rebecca M. Lennen, Chemical and Biological Engineering, University of Wisconsin-Madison, Madison, WI and **Brian F. Pfleger**, Chemical and Biological Engineering, University of Wisconsin Madison, Madison, WI

Finding a sustainable alternative for today's petrochemical industry is a major challenge facing chemical engineers and society at large. To be sustainable, routes for converting solar energy into organic compounds for use as both fuels and chemical building blocks must be identified, understood, and engineered. Advances in synthetic biology and other biological engineering disciplines have expanded the scope of what can be produced in a living organism. As in other engineering disciplines, synthetic biologists want to apply a general understanding of biology to construct complex systems from well-characterized parts. Once novel synthetic biological systems (e.g. enzymes for biofuel synthesis) are constructed, they must be engineered to function inside living cells without negatively impacting the host's physiology. In most cases first generation systems fail to meet this goal. My group uses systems biology tools to identify metabolic, regulatory, and/or physiological barriers which often can be overcome with metabolic engineering strategies. In this talk, I will describe how we have combined synthetic biology and systems biology to develop strains of bacteria that produce compounds derived from fatty-acids starting from sustainable feedstocks to engineering efforts. Our work has combined functional genomics analysis, synthetic biology construction techniques, bioinformatics, and metabolic modeling to metabolically engineer superior microorganisms. We have tested our strains in media formulated with biomass-derived sugars and are developing photosynthetic microorganisms to by-pass the biomass middle man.

(175f) Repressor-Based Tools for Cell-Cell Communication in Synthetic Biology

Jasmine Shong and **Cynthia H. Collins**, Chemical and Biological Engineering, Rensselaer Polytechnic Institute, Troy, NY

A grand challenge in synthetic biology is the development of novel intercellular signaling systems. We have engineered the acyl-homoserine lactone (AHL)-dependent transcriptional repressor, EsaR, and the *esa* promoter for use in synthetic microbial communities. The availability of intercellular communication-dependent repressors will enable new circuits and network behaviors not feasible with existing parts (e.g., the transcriptional activator LuxR). We have used directed evolution to identify a number of EsaR variants that respond to AHL concentrations between 5 and 10,000 nM. We have also engineered new EsaR dependent promoters. Here we have taken advantage of the AHL-dependent DNA-binding activity of EsaR to build synthetic AND gates that recognize both an extracellular signal and cell density. We have characterized our multi-input promoters and have found that activity can be tuned by varying AHL production, AHL sensitivity and addition of an endogenous inducer molecule. These new transcriptional repressors and promoters will enable construction of novel engineered density-dependent or multicellular systems for metabolic engineering applications.

(175g) A Platform for Synthetic, Orthogonal, Intercellular Communication Via Synthetikines

Rachel M. Dudek[1], Nichole Daringer[1] and Joshua N. Leonard[2], (1)Chemical and Biological Engineering, Northwestern University, Evanston, IL, (2)Chemical and Biological Engineering, Chemistry of Life Processes Institute, R.H. Lurie Comprehensive Cancer Center, Northwestern University, Evanston, IL

A technology enabling the transmission of independent, synthetic, high information content signals between cells in a multicellular network remains a grand challenge in mammalian synthetic biology. This capability could enable the construction of networks involving division of labor, spatial organization, and coordinated functions eventually leading to applications including regenerative medicine and tissue engineering, as well as new tools for fundamental research. Protein-based signaling mediators are well-suited to this role, and molecules such as cytokines and chemokines serve exactly this function in natural regulation of multicellular organisms. However, to date we lack synthetic biology ligand-receptor platforms suitable to sending and receiving protein-based signals. To meet this need, we have developed a novel orthogonal signaling platform comprising modular synthetic cytokines (which we have termed, "synthetikines") and corresponding modular synthetic receptors that can detect and discriminate between these signaling molecules. We show how modular receptor design enables both the use of orthogonal signal transduction mechanisms and optimization of desirable receptor performance characteristics. We have devised synthetikines that enable construction of both paracrine signaling (i.e., via secretion of soluble ligands) and juxtacrine signaling (i.e., by cell surface-bound ligands), each of which play key roles in processes such as homeostasis and development. Here, we characterize a foundational platform based on mammalian cells, although the core technology could also be adapted to enable interkingdom signaling (e.g., for interrogating or manipulating interactions between host and bacteria in the gut). This novel capability in the mammalian synthetic biology toolkit should prove useful for both investigating and implementing design principles for engineering synthetic multicellular networks.

(175h) Refactoring the Partial Photosynthesis Gene Cluster for the Biosynthesis of the Key Pigment, Bacteriochlorophyll a

Hui Zhou, Department of Biological Engineering, Massachusetts Institute of Technology, Cambridge, MA, J. Thomas Beatty, Dept. of Microbiology & Immunology, University of British Columbia, Vancouver, BC, Canada and Christopher A. Voigt, Synthetic Biology Center, Department of Biological Engineering, Massachusetts Institute of Technology, Cambridge, MA

In prokaryotes, the genes responsible for complex biological functions are often physically clustered on the chromosome for co-regulation. As a goal for synthetic biology, it is important to design and build manipulable gene clusters for the further construction of biological systems with higher hierarchy and useful biological functions. In order to achieve this goal, microbial photosynthesis machinery was chosen as a target for refactoring. Taking the purple non-sulfur bacterium *Rhodobacter sphaeroides 2.4.1* as an example, almost all the genes responsible for photosynthesis are clustered on the chromosome spanning in an approximately 40 kb region. We are using a systematic approach to re-design this photosynthesis gene cluster by rebuilding it from the bottom-up using synthetic and well-characterized genetic parts. Focusing on the biosynthetic operons for the key pigment bacteriochlorophyll *a* (BChl *a*) biosynthesis, all the essential genes will be chosen for knock-out and complementation, with the exception of regulatory genes. An architecture with mono-cistronic expression of each gene will be adopted for the construction of the synthetic gene cluster. Each gene is being designed to be under the control of tunable genetic parts, including promoter, ribosome binding site, transcription terminator and insulator. The codons for the genes are adjusted to be as different as possible from the wild type genes, thus excluding possible internal regulation sequences. Finally, a simplified environment sensor will be implemented in a genetic circuit to regulate the gene expression profile. The synthetic gene cluster for the biosynthesis of BChl *a* may provide a modular platform for further combinatorial biosynthesis of BChl *a* derivatives, which may expand the chemical space for additional and novel photosynthetic pigments. Via this refactoring process, this gene cluster may also be easily transferred to other non-photosynthetic microorganisms, facilitating the engineering of artificial photosystems.

(239a) Proteomic Study of High Vaccine DNA Producer

Jonathan Meade, Chemical Engineering, Carnegie Mellon, Pittsburgh, PA, Patrick R. Bartlow, Bioengineering, University of Pittsburgh, Pittsburgh, PA, Ram N. Trivedi, Chemical and Petroleum Engineering; Microbiology and Molecular Genetics, University of Pittsburgh, Pittsburgh, PA, Saleem A. Khan, Microbiology and Molecular Genetics, University of Pittsburgh, Pittsburgh, PA, Mohammad M. Ataai, Chemical and Petroleum Engineering, University of Pittsburgh, Pittsburgh, PA and Michael M. Domach, Chemical Engineering, Carnegie Mellon University, Pittsburgh, PA

Plasmid DNA is historically perceived as a means for recombinant protein production. However, increasingly it has become recognized as a therapeutic product in its own right. With many DNA vaccine clinical trials now underway and some showing promise, large quantities of plasmid DNA could be needed in the future. In addition to enabling the therapeutic applications, producing plasmid DNA at a high yield now could improve the throughput of basic experimentation that requires transfection as a step.

The goal of the work underway is to develop a production platform that has high volumetric productivity of plasmid DNA when inexpensive and simple minimal media is used. We previously reported the use of *E. coli* DH5α and a mutated plasmid that achieved copy numbers in excess of 10,000. In this prior work, plasmid-containing cells were selected for by sucrose and one facet of the mutation scheme was to alter the origin of replication region in order to reduce RNA1/RNA2 interaction. Apart from the high copy numbers, there were two interesting findings: (i) the growth rate was not affected as much as one would expect and (ii) a drastic alteration to cell morphology occurred.

To resolve how the host adapts to the sizable burden imposed, proteomic and metabolic studies have commenced, and this presentation will discuss some of the findings. The changes that occur in the relative abundance of many ribosomal proteins are of particular interest. These changes suggest that adaptation may be occurring in the high DNA-producers at a more general level where ribosomal assembly and recycling is altered. Such changes and others may also account for the altered cell morphology. In addition to proteomics, the effect of different carbon sources on product yield will be discussed as well as the changes in metabolic enzymes and stress proteins that also correlate to more traditional measurements such as acetate analysis.

(239b) Metabolic Tradeoffs in Resource Investment and Function: A Switch or a Dial?

James P. Folsom, **Reed Taffs** and Ross P. Carlson, Chemical and Biological Engineering, Montana State University, Bozeman, MT

An economics-based analysis of stoichiometric metabolic models quantified a continuum of molecular-level, resource allocation tradeoff strategies that permit competitive cellular functioning under continuous gradients of nutrient scarcities. The theoretical biochemical pathway resource requirements differ between distinct nutrient limitations as well as across the severity of the considered stress. The approach provides attractive ecological explanations for a variety of well-known physiological responses to specific culturing conditions, as well as the ability to extract a relatively small number of biologically important flux distributions from a stoichiometric model's enormous steady-state solution set. The reduced solution space provides biological insights on better fits of a variety of quantitative flux measurements than are possible using other published cellular objective functions, but published data for further testing the graduated nature of these predictions is surprisingly lacking. The presented work fills some of that gap, generating and examining the experimental proteomic and physiological responses of *Escherichia coli* to four levels of iron, nitrogen, and carbon restriction. The results provide a theoretical basis for the relative strength of evolutionary selection for optimality as well as metabolic simplicity, and additionally suggest alternative approaches to the redirection of metabolism for industrial applications.

(239d) Does the Introduction of Non-Native Secondary Metabolite Pathways Affect Flux in Upstream Primary Metabolism? A 13C-Metabolic Flux Analysis of Yeast Engineered to Produce Artemisinin Precursors

Matthew Conway[1], Ashish Misra[1], Eddy Agbo[2] and Ganesh Sriram[1], (1)Chemical and Biomolecular Engineering, University of Maryland, College Park, MD, (2)Fyodor Biotechnologies, Baltimore, MD

Does the Introduction of Non-Native Secondary Metabolite Pathways Affect Flux in Upstream Primary Metabolism? A 13C-Metabolic Flux Analysis of Yeast Engineered to Produce Artemisinin Precursors

There is rising interest in metabolic engineering of microorganisms such as yeast for production of valuable non-native compounds. The successful genetic engineering of a non-native pathway may have systems-level implications on other, apparently unrelated pathways. 'Omics tools permit an investigation of such phenomena. Specifically, isotope-assisted metabolic flux analysis enables the quantitative visualization of fluxes through metabolic pathways in a systemwide manner and thereby enables identification of factors contributing to lack of success of a genetic engineering attempt. In this study, we employ ^{13}C isotope-assisted metabolic flux analysis for investigating fluxes in *Saccharomyces cerevisiae* strains engineered to express genes in downstream, secondary metabolie pathways in an inducible manner. This presentation will report the results of this analysis, specifically how such secondary pathways can potentially influence primary metabolic pathways.

(239e) Metabolomics and Integrated "Omics" Analysis of the Mesenchymal-Epithelial Transition in Ovarian Cancer

Kathleen Vermeersch[1], Lijuan Wang[2], John McDonald[2] and Mark P. Styczynski[1], (1)School of Chemical & Biomolecular Engineering, Georgia Institute of Technology, Atlanta, GA, (2)School of Biology, Georgia Institute of Technology, Atlanta, GA

While there is a growing general awareness of the importance of metabolic changes in the onset and progression of cancer, the molecular basis and significance of these metabolic changes are only beginning to be understood. One aspect of cancer that is particularly poorly characterized - on the metabolic level as well as more generally - are transitions between epithelial and mesenchymal phenotypes. The epithelial-mesenchymal transition (EMT) is the process by which cells lose their epithelial characteristics and take on more mobile, mesenchymal-like phenotypes. MET is the reverse process, where cells convert from mesenchymal to epithelial phenotypes. Mesenchymal-like cancer cells are more migratory than epithelial-like cancer cells, more invasive, and more resistant to chemotherapy. EMT/MET are normal processes during organismal development, but they are hijacked by cancer cells, resulting in cells leaving the primary tumor site (EMT) and migrating to establish distant sites of tumor growth (MET). The mechanisms underlying these processes are also believed to play a role in conferring "cancer-initiating cell" (CIC) characteristics: stem cell-like properties including the potential to differentiate into the many cell types present in tumors. Thus, EMT/MET are the underlying processes necessary for metastatic cells to travel to and successfully colonize distant sites. While the molecular pathways of EMT/MET are still being uncovered, the metabolism of these processes has yet to be studied.

Recently, we created a novel ovarian cancer model of EMT/MET, processes that are difficult to observe or control *in vivo* or *in vitro*. Using HEY cells, which have a mesenchymal-like phenotype, we found that overexpression of the microRNA miR-429 results in MET[1]. This is the only existing *in vitro*miRNA-controlled model of ovarian cancer EMT/MET. In this work, we seek a multi-level "omics" characterization of MET. During miRNA-induced MET and susbsequent EMT reversion over the course of six days, samples are taken for metabolomics, proteomics, and transcriptomics analysis. Metabolomics analysis was done using two-dimensional gas chromatography coupled to mass spectrometry (GCxGC-MS) on mostly polar small molecule analytes.

We have found significant differences across multiple time points between MET and control samples based on both intracellular and extracellular metabolomics analyses. These results suggest that there is a distinct metabolic phenotype for epithelial and mesenchymal phenotypes; integration of proteomic and transcriptional data is being used to more fully understand these distinct phenotypes and to help identify whether there are metabolic aspects of MET and EMT that can be targeted for potential cancer treatments.

Reference:

[1] Chen J, Wang L, Matyunina LV, Hill CG, McDonald JF. Overexpression of miR-429 induces mesenchymal-to-epithelial transition (MET) in metastatic ovarian cancer cells. *Gyn Oncol*. 2011;121:200-5.

(239f) Multiplex, Longitudinal Immune Cellular Functional Proteomic Analyses Correlated with the Clinical Outcome of a Melanoma ACT Immunotherapy Trial

Chao Ma[1], Ann Cheung[1], Antoni Ribas[2] and James Heath[1], (1)California Institute of Technology, Pasadena, CA, (2)University of California Los Angeles, Los Angeles, CA

Background: Adoptive cell transfer (ACT) of transgenic T lymphocytes expressing tumor antigen specific T cell receptor (TCR) can mediate objective responses in 30%-50% of metastatic melanoma patients and result in long term tumor regression. However, little is known about the fate of engineered T cells once transferred and the dynamics of host immunity, which is important for understanding the biological mechanism of the therapy and differing long term efficacy between patients.

Methods: Utilizing newly developed, highly multiplexed clinical immune diagnostics methods (Ma, C. *et al*. Nature Medicine 17, 738-743 (2011)), we designed a comprehensive study that monitored the engineered and acquired immune response in 8 metastatic melanoma patients over 8-10 time points in the context of an ACT therapy with transgenic TCR T cells. (1) By integrating a clinical microchip platform with florescent activated cell sorting, we quantified 19 functional proteins secreted from single cells to evaluate the functional proteomics of 6 phenotypically defined T cell subsets; (2) We enumerated the abundance of 35 melanoma antigen specific T cell subsets using a peptide/MHC tetramer microarray; (3) We measured a panel of 37 blood proteins that incorporates melanoma associated markers, cytokines and chemokines by an ELISA-like assay. These assays yielded millions of high dimensional and highly informative readouts and we developed high dimensional machine learning algorithm to analyze the data.

Results: In all patients, there was an initial clinical antitumor response followed by disease progression. The change of blood marker abundance reflected this general process, starting from lymphodepletion pre-conditioning, immune system recovery, immune response, to tumor regression. This clinical course correlated with changes observed in T cell functional proteomics. The transgenic MART-1 T cells expanded rapidly after infusion, producing multiple cytotoxic and antitumor cytokines. However, their abundance and polyfunctionality diminished within 2-3 weeks. At around 30 days post infusion, a multi-clonal expansion of melanoma specific CTLs emerged in patients with regressing metastatic lesions. This wave of acquired endogenous cellular immunity exhibited antitumor functions initially, but gained inhibitory and non-specific functions and diminished when tumor started to progress. The patient with the most endogenous melanoma specific T cell populations emerged and with the strongest functions in the *in vivo* acquired immune response showed the most effective long term tumor control; weaker second wave response was associated with weaker therapeutic efficacy.

Conclusions: The melanoma-specific T cells within the ACT went through a process of *in vivo* functional changes that correlated with clinical antitumor activity. The appearance and the persistence of the *in vivo* acquired immune response featured by engineered and newly appeared endogenous antitumor cellular immunity provided indications of long term therapeutic efficacy. Therefore, the study points to the need to develop combinatorial therapeutic approaches that could maintain T cell functionality to improve the efficacy of ACT therapy.

(239g) Phospho-Proteomics Reveals a Metabolic and Signaling Amplification Loop Leading to Cell Death Following Glucose Deprivation

Nicholas A. Graham[1], Martik Tahmasian[1], Bitika Kohli[1], Evangelia Komisopoulou[1], Maggie Zhu[1], Igor Vivanco[2], Michael A. Teitell[3], Hong Wu[1], Antoni Ribas[4], Roger S. Lo[5], Ingo K. Mellinghoff[2], Paul S. Mischel[3] and Thomas G. Graeber[1], (1)Medical and Molecular Pharmacology, University of California, Los Angeles, Los Angeles, CA, (2)Memorial Sloan-Kettering Cancer Center, New York, NY, (3)Pathology and Laboratory Medicine, University of California, Los Angeles, Los Angeles, CA, (4)Hematology and Oncology, University of California, Los Angeles, Los Angeles, CA, (5)Dermatology, University of California, Los Angeles, Los Angeles, CA

Introduction: Tumor cells exhibit an altered metabolism compared to non-transformed cells, consuming glucose and glutamine and producing lactate at prodigious rates. One consequence of deregulated tumor metabolism is that cells can become dependent on the supply of metabolic substrates such as glucose or glutamine for viability. Based on evidence that oncogenic signaling can impose this metabolic inflexibility (Choo et al, 2010), as well as the concept that complex networks link metabolism and signal transduction (Wellen et al, 2010), we conducted a systems level investigation into the role of signal transduction underlying cell death following metabolic perturbation in cell systems dependent on glucose for survival.

Materials and Methods: Phospho-tyrosine signaling and viability of U87 glioblastoma cells were assessed by Western blotting and trypan blue exclusion, respectively. Quantitative, label-free mass spectrometry was performed as previously described (Rubbi et al, 2011) on peptides immunoprecipitated using a pan-specific anti-phosphotyrosine antibody (clone 4G10, Millipore). Reactive oxygen species were measured by flow cytometry using DCF-DA. Protein tyrosine phosphatase activity was measured using a substrate dephosphorylation assay.

Results and Discussion: In cells dependent upon glucose for survival, glucose withdrawal rapidly induces supra-physiological levels of phospho-tyrosine signaling. Increased phospho-tyrosine signaling was not observed in glucose withdrawal-resistant cells. Using unbiased mass spectrometry-based phospho-proteomics, we found that glucose withdrawal initiates a unique signature of phospho-tyrosine activation compared to other stimuli (eg, growth factor stimulation, the phosphatase inhibitor vanadate). Building upon the observation, we demonstrate that glucose withdrawal activates a positive feedback loop whereby reactive oxygen species (ROS) oxidize and inhibit protein tyrosine phosphatases (PTPs), leading to increased tyrosine kinase signaling, which induces further ROS generation. Following glucose deprivation, ROS generation and tyrosine kinase signaling synergize to amplify ROS levels until cells undergo ROS-mediated cell death.

Conclusions: We demonstrate that a metabolic perturbation (ie, glucose starvation) provokes a signaling-based positive feedback loop that amplifies ROS levels until cells surpass a toxicity threshold. This positive feedback loop demonstrates the complex, systems level integration of homeostatic control mechanisms for metabolism (eg, redox balance) and tyrosine kinase signaling (eg, PTPs). Furthermore, this systems integration offers a scaffold for synergistic combinations of therapeutics targeting signaling and metabolism.

References: Choo et al, *Mol Cell* (2010); 38: 487-499.

Rubbi et al, *Sci Signal* (2011); 4: ra18.

Wellen et al, *Genes Dev* (2010); 24: 2784-2799.

(239h) Cooperation Between Subtypes of Tumor Cells Promotes Malignancy Studied by Multiplexed Super-Resolution Single Cell Proteomic Assay

Jun Wang, Chemistry, California Insitute of Technology, Pasadena, CA and James Heath, Chemistry and Chemical Engineering, California Institute of Technology, Pasadena, CA

Glioblastoma multiforme (GBM) is the most common and malignant brain cancer in adult. Amplification of epidermal growth factor receptor (EGFR) expression has been found in ~50% of primary GBMs, and associated with poor prognosis. The EGFR variant, EGFRvIII, represents the most common mutant and persistently activates cancer cells to be highly diffusive. Xenograft assay reveals that, although $EGFRvIII^{+}$ cancer cells grow much faster than $EGFRvIII^{-}$ cells in vitro, those two cell subtypes proliferate in a manner similar to symbiosis in vivo. Using our single cell barcode chip, we captured both secreted proteins and intracellular phosphoproteins at the single cell level when two cells of different phenotypes were cultured in an artificial microenvironment. Our technology enables quantification of >8 proteins of a single cell with a resolution of ~50 proteins per cell, and gathers information from ~1000 cell pairs. Using information theory and Le Chatelier's principle, the interaction between cell pairs is predicted, which also explains the activation of $EGFRvIII^{-}$ cells and fast proliferation in tumor. We construct a model in statistic mechanics to simulate cell subtype composition in a tumor. We further investigate the interaction of $EGFRvIII^{+/-}$ cells and stromal cells as well as immune cells in our microplatform. The understanding of tumor-host cell interaction will give insight into the tumor heterogeneity and complicated tumor social network.

(254a) Integrated Processing of Microalgae for Valuable Food Ingredients, Chemicals and Fuel Production

Kaige Wang[1], Yi Liang[2], Zhiyou Wen[2] and Robert Brown[1], (1)Center for Sustainable Environmental Technologies, Iowa State University, Ames, IA, (2)Food Science and Human Nutrition, Iowa State University, Ames, IA

Lipid-based biofuel production by microalgae has captured interest of researchers due to microalgae's high biomass yield and ability to synthesize and accumulate large quantities of lipid. However, high cost of algal cultivation and processing has limited the commercialization of algal fuel production. Developing a microalgae biorefinery strategy, i.e., using every components of algal biomass for fuels and chemicals production is a feasible approach for commercially utilization of the algal biomass.

This study reports a hybrid processing for treating the microalgae *Schizochytrium limacinum*. Crude glycerol, a major byproduct in biodiesel production plants, was used as substrate for producing high content of algal oil containing DHA (22:6, n-3), a high value omega-3 polyunsaturated fatty acid with beneficial effects on human health. The residues after oil-extraction from *S. limacinum* were then catalytically pyrolyzed on a micro-furnace pyrolyzer with zeolite catalyst being used. Aromatics and ammonium was generated from this catalytic pyrolysis process. Among the aromatics, benzene, toluene and xylene, which are valuable basic petrochemicals,were the dominant ones. Most of the nitrogen in the residues was released in ammonium form, which may be in turn used as nutrient for algal culture. In summary, all the components in the biomass of the *S. limacinum* were utilized for producing chemicals, fuels and value-added products which can be future enhanced the economic for algal industry.

(254b) Pyrolytic Fractionation of Oleaginous Feedstocks

Balakrishna Maddi, Sridhar Viamajala and Sasidhar Varanasi, Chemical & Environmental Engineering, University of Toledo, Toledo, OH

Traditional pyrolysis of oleaginous biomass results in bio-oils that contain a mixture of compounds derived from protein, starch and lipids – constitutive biopolymers of these feedstocks. As a result, such biooils have to undergo energy-intensive denitrogenation and deoxygenation to render them suitable as fuels. We have recently developed a "*pyrolytic fractionation*" technique based on the observation that the protein, starch and lipid portions undergo thermal degradation over distinct non-overlapping temperature intervals. Proteins are most thermally labile followed by starch and lipids. Hence, conducting the pyrolysis in three temperature stages provides an opportunity to recover homogenous and biopolymer-specific biooils. Bio-oils from protein and starch can find other applications, in addition being lower grade fuels. For instance, N-rich biooil from protein can serve as a source of high value N-compounds and levoglucosans from starch pyrolysis could serve as precursors for value-added polymers. Similarly, bio-oil from lipid fraction, predominantly containing free fatty acids, can be converted to fuels (such as biodiesel or hydrocarbons) or other higher value fatty acid derivatives. Finally, the bio-char can be a soil amendment.

In this presentation, results showing component-specific biooil compositions and yields from pyrolytic fractionation of algal biomass and oil seeds will be presented. We will also present a comparison of energy needs for pyrolytic fractionation relative to other themochemical methods for conversion of oleaginous feedstocks.

(254c) Hydrothermal Carbonization and Supercritical Ethanol in Situ Transesterification for the Production of Algal Biodiesel

Robert Levine, Chemical Engineering, University of Michigan, Ann Arbor, MI and Phillip E. Savage, Chemical Engineering, University of Michigan, Ann Arbor, Ann Arbor, MI

The production of transportation fuels from microalgae is complicated by the fact that algae are typically grown suspended in a dilute aqueous media and have a high demand for nitrogen and phosphorous. In this work, we explore a new and environmentally preferable strategy for algal biofuel production. There are three major components of the bio-refinery model we have developed: 1) cultivation of oil-rich algae using recycled water and nutrients; 2) treatment of the harvested algae through hydrothermal carbonization to produce a biochar; and 3) conversion of the carbonized algal biomass into transportation fuels using in situ transesterification with supercritical ethanol.

Lipid-rich *Chlorella* was grown in bioreactors until it contained approximately 50% oil, then carbonized for 15 to 90 min at 180 to 235 °C. Biochar obtained by filtration retained >90% of lipids of the original biomass and was converted into biodiesel by reacting the solids with ethanol at 270-325 °C for 15-180 min. Fatty acid ethyl yields in excess of 90% were obtained while employing 10-100 times less ethanol than previously reported for comparable transesterification reactions. The nutrients liberated during hydrothermal carbonization were used to produce additional algal biomass, thereby recycling N and P. Overall, our work suggests that it is possible to produce net-energy positive algal biofuels with significantly less environmental impact than the petroleum fuels they replace.

(254d) Integrating Photo-Bioreactor and Fermentor to Produce Biofuels and Bioelectricity

Alim Dewan, Marci Kerls and M. Nazmul Karim, Chemical Engineering, Texas Tech University, Lubbock, TX

We have investigated simultaneous production of biofuels and bioelectricity from an integrated bioreactor that combines photo-synthesis and fermentation. In the integrated bioreactor, a fermentor and a photo-bioreactor were combined in a way that the carbon dioxide produced in the fermentation can be used to grow algae in the photo-bioreactor. Also, the reactors were attached to a proton exchange membrane in between so that they form a biological fuel cell. To synchronize the carbon dioxide production in the fermentation and the carbon dioxide utilization in photo-synthesis, the fermentor was run continuously but the photo-bioreactor was run in batch mode. An autotrophic microalgae, *Chlorella vulgaris,* and a bear yeast, *Saccharomyces bayanus,* were grown in the photo-bioreactor and in the fermentor, respectively. The ethanol production by the yeast, oil production by the algae, and electricity generation were monitored over time. The electricity generation was characterized by potentiodynamic polarization. In a two-liter size integrated bioreactor, we observed an ethanol productivity of 0.49±0.37 g l^{-1} hr^{-1}, algae oil productivity 2.5 mg l^{-1} hr^{-1}, power density of 70.61 ± 12.41 mW/m^2 of the electrode, and current density of 155.02 ± 31.23 mA/m^2. Preliminary scale-up studies show a great potential for this integrated system.

(254e) Determination of Lipid Hydrolysis Kinetics in Soybean Oil and Algal Systems

Joshua C. Wissinger[1], Amber L. Bosley[1] and Constance Schall[2], (1)Chemical and Environmental Engineering, University of Toledo, Toledo, OH, (2)Chemical & Environmental Engineering, University of Toledo, Toledo, OH

Determination of lipid hydrolysis kinetics in soybean oil and algal systems

Joshua C. Wissinger, Amber L. Bosley, Constance A. Schall

The third generation of biofuels, algae, is of growing interest as a renewable feedstock due to its high yield of energy per unit area, use of carbon dioxide and other available resources for growth, and minimal competition with the food industry. Biodiesel and jet fuel can be produced from lipid-rich algae. However, extracting lipids from algae is complicated due to the high water content of harvested and mechanically dewatered algae (60 to 90% water). Thus, there is a need for processes that convert wet algal biomass into fuels or fuel precursors directly in the aqueous phase. Such a process alternative is to use the water in the harvested and partially dewatered algae for hydrolysis of lipids to free fatty acids (FFA), forming aqueous and oil rich phases for ease of extraction and separation of the lipid component. Lipid processing techniques include refining FFA by catalytic cracking and hydrotreating for jet fuel production, or the UOP/Eni Ecofining™ process to obtain clean green diesel. In our first step, subcritical water is reacted with lipids producing FFA by lipid hydrolysis eliminating the need for expensive and water sensitive catalysts. In the initial development of an empirical kinetic model, soybean oil, a system much less complex than alga, was used in hydrolysis studies at subcritical water (SCW) conditions. This model was applied to algal lipid hydrolysis. The robustness of our model has been tested by verifying both the autocatalytic behavior of lipid hydrolysis reactions, as well as hydrolysis reactions over a range of temperatures.

(254f) Microwave-Assisted Subcritical Water Extraction of Lipids From Wet Algae

Harvind Kumar Reddy[1], Yingqiang Sun[1], Yin Li[2], Tapaswy Muppaneni[1], Sundaravadivelnathan Ponnusamy[1], Shuguang Deng[1], Tanner Schaub[3], Barry Dungan[4], Francisco Holguin[4], Peter Lammers[5], Wayne Voorhies[6] and Peter Cooke[7], (1)Chemical Engineering, New Mexico State University, Las Cruces, NM, (2)Chemical and Biological Engineering, Zhejiang University, Hangzhou, China, (3)College of Agricultural, Consumer and Environmental Sciences, New Mexico State University, Las Cruces, NM, (4)College of Agricultural, Consumer and Environmental Sciences, New Mexico State university, Las Cruces, NM, (5)Energy Research Laboratory, New Mexico State University, Las Cruces, NM, (6)Molecular Biology, New Mexico State university, Las Cruces, NM, (7)Core University Research Resource Laboratory, New Mexico State University, Las Cruces, NM

Microwave-assisted extraction of algal oil at subcritical water conditions was demonstrated with wet algae samples containing up to 90% of water. The extraction efficiency of this method was compared to those of the conventional solvent extraction methods, Folch and Bligh & Dyer. The extraction parameters including biomass loading, temperature and extraction time are optimized after executing a set of designed experiments. Algal oil was produced after purification of bio crude with activated charcoal. The residual algal biomass was also analyzed for its fatty acid and energy content to validate potential byproduct development as part of algal biorefinery. Energy calculations were performed for microwave assisted subcritical water extraction and compared with those of the conventional methods. Analytical instruments including GC-MS, FT-ICR were used to analyze the biomass and biocrude. Thermogravimetric analysis (TGA), TEM and SEM equipped with EDS were used to characterize the biomass in different stages of the extraction process.

(295a) Multi-Scale Agent-Based Modeling of Cancer Cell Chemotaxis within a Microfluidic Assay: Investigating the Role of Receptor Dynamics and Ligand Isoforms

S. Laura Chang[1], Stephen P. Cavnar[2], Danielle Trakimas[1], Shuichi Takayama[2], Gary D. Luker[3,4] and Jennifer J. Linderman[1], (1)Department of Chemical Engineering, University of Michigan, Ann Arbor, MI, (2)Department of Biomedical Engineering, University of Michigan, Ann Arbor, MI, (3)Department of Microbiology and Immunology, University of Michigan, Ann Arbor, MI, (4)Department of Radiology, University of Michigan, Ann Arbor, MI

Chemotaxis is proposed to be a key mechanism that allows cancer cells to invade surrounding tissue and vasculature, and also to extravasate into distant locations leading to metastasis. The signaling axis comprised of the chemokine CXCL12 and its receptor CXCR4 has been implicated as a major chemotactic proponent in breast cancer and more than 20 other cancers. Inhibition of CXCR4 has not resulted in clinically successful cancer therapy, prompting the search for other targets. A second receptor, CXCR7, is also present in the breast tumor environment and can scavenge CXCL12. Understanding how CXCR7 can shape ligand gradients on a molecular level, and influence cancer cell chemotaxis within the tissue scale, is unclear. In addition, CXCL12 exists as multiple isoforms, yet studies probing its role in cancer have only focused on one isoform. These isoforms have varying affinities for cell surface receptors and glycosaminoglycans, as well as experimental device surfaces, ultimately changing the presentation of the ligand to the receptor. These complex interactions are overlooked in standard non-physiologic cell culture models and confound our understanding of chemotactic gradient formation, sensing, and its role in cancer.

In order to investigate the molecular mechanisms that regulate gradient formation and cancer cell responses to these gradients, we constructed a data-driven multi-scale agent-based model that simulates cancer cell chemotaxis within a microfluidic assay. First, we use *in vitro* data to construct an ordinary differential model to describe ligand and receptor binding, internalization, recycling and degradation for both receptors. Second, we take these single-cell equations and embed them within an agent-based model describing how cells interact and chemotax within a microfluidic device. Thus the molecular scale behavior drives the cellular level decisions, which ultimately dictate the overall organization of the cells in the device. We build and validate the model based on data in Torisawa et al. (*Integrative Biology* 2:680-686, 2010). By performing uncertainty and sensitivity analyses with Latin Hypercube Sampling and Partial Rank Correlation Coefficients, we then identify and analyze processes that have the greatest influence on chemotaxis. We find that ligand and receptor dynamics play a large role in shaping gradients and thus modulating cell behavior. By incorporating the partitioning of CXCL12 from its soluble state to its glycosaminoglycan-, surface- and receptor-bound states in the molecular scale of our model, we can also investigate how different CXCL12 isoforms affect the gradient, concentration, and cell behavior. Overall, our model is able to provide information on how molecular- and cellular-scale events determine cellular organization in a microfluidic device, and can be extended to explore cell behavior in tissue in response to CXCL12.

(295b) In Silico Prediction of Cancer Mechanism of Action

E. A. Whitebay, J. D. Ramsey, B. J. Neely and K. A. M. Gasem, School of Chemical Engineering, Oklahoma State University, Stillwater, OK

Currently cancer is the second leading cause of death in the US and is projected to become the leading cause in the near future. Traditional chemotherapy remains a key treatment method, offering distinct advantages over other treatment options, especially in the treatment of metastasized tumors. To develop new chemotherapeutic drugs, greater knowledge concerning the mechanisms of action (MoA) of current cancer drugs is required. By understanding how current drugs inhibit cancer growth through their underlying MoA and what molecular features affect the MoA of a drug, we can develop new and better chemotherapeutic agents.

Quantitative structure-activity relationship (QSAR) prediction of MoA allows us to relate structural features to the behavior of a chemical, thus enabling greater insight into the relationship between drug activity and molecular structure. Further, various types of cancer demonstrate varying responses to different MoAs. Classifying new potential drugs by MoA can reduce experimental testing by restricting testing candidates to only those potential drug candidates predicted to have a specified level of effectiveness. Previous research has employed QSAR predictions in an effort to predict the MoA of new, untested chemicals; however, these predictions required experimental input, usually involving the growth inhibition profiles for a panel of cancer cells lines. Our efforts have centered on advancing beyond the current state of the art by providing QSAR predictions that are truly *a priori* in nature. Without any experimental knowledge, we can classify the National Cancer Institute Anti-cancer Agent Mechanism Database, which consists of 122 molecules. Our model has an overall predictive accuracy of 84%, with 10% of the molecules not classified into any MoA class and 6% of the molecules classified into an incorrect class.

(295c) Linking Microbial Structure to Function in Representative Simulated Biological Systems

Ian Marcus, Chemical and Environmental Engineering, University of California, Riverside, Riverside, CA and Sharon L. Walker, Department of Chemical and Environmental Engineering, University of California, Riverside, Riverside, CA

Linking Microbial Structure to Function in Representative Simulated Biological Systems

Ian Marcus, PhD Candidate

Professor Sharon L. Walker, Advisor

Chemical and Environmental Engineering, University of California, Riverside

The common paradigm when studying pathogenic microorganisms in the lab has been to grow the cells as a single strain in rich nutrient media and then to harvest and evaluate the organisms' phenotypic and/or genotypic characteristics. However, this traditional approach overlooks a critical fact that these pathogens do not exist in such idealized conditions. Microorganisms survive in complex communities of multiple species of bacteria, archaea, fungi, and protozoa, though bacteria make up most of the biomass. It is therefore imperative to study the microbial community along with the pathogens as a biological system, and to establish the community's effect on the subsequent introduction of these pathogens into the environment. Ultimately with this knowledge, the impact of the host microbial community on the virulence of pathogens in aquatic environments can be identified and the potential environmental health risk assessed.

The overall goal of this investigation is to ascertain the effect the addition of pathogenic *E. coli* O157:H7 has on the host microbial community through commonly encountered biological systems relevant to water quality. A series of experiments have been used to evaluate the microbial community structure and function in three simulated systems: a colon, septic tank, and groundwater. Pyrosequencing was performed to determine the microorganisms present, extensive physical-chemical analyses on the community of microorganisms and pathogen (i.e. electrophoretic mobility, cell size, surface charge density, hydrophobicity and extracellular polymeric substances) have been performed in each of these systems, as well as transport experiments of the community and pathogen in a macroscopic, packed-bed column system to simulate their behavior once introduced into the environment. The results of this ongoing study will give greater insight into the understanding of the way in which environmental factors influence the fate and transport of pathogenic microorganisms when they enter and are transported in aquatic environments. Work to date has shown that the physical-chemical characteristics and transport of the microbial community change depending on exposure between the model colon and representative aquatic systems. Specifically, the microbial community is more hydrophilic, has more negative electrophoretic mobility, lower surface charge density, a higher quantity of EPS and has lower attachment efficiency when sampled from the model colon, as compared to the representative aquatic systems. With the addition of the model pathogen the hydrophobicity and electrophoretic mobility increases, while other trends remain consistent. The research presented will include comparisons of the fate and transport of the microbial community with the microorganisms present to link the microbial structure to function in synthetic biological systems.

(295d) Nanoscale to Macroscale Modeling of Biochemical Linked Muscle Response

Richard Long Jr., Chemical Engineering, NMSU, Las Cruces, NM

The Huxley model of cross-bridge filament interaction allows modeling at the nanoscale. By integrating across the entire sarcomere it is possible to compute the response at the macroscale. By knowing the biochemistry of the excitation-contraction coupling, it is possible to compute the nano scale and macroscale mechanical response action for the muscle. By structural models of the filaments, it is possible to compute the multidimensional response of the structural features of the muscle tissue extant. The response of the S! fragment of the actomyosin crossbridge, enables one to compute the nanoscale force model of the response. By combining the linear momentum balance with the chemical rate equations in the biochemical cycle, it is possible to compute the macroscale force balance. The nanoscale micro-environment shows the effect of the small molecule response on the filament action in the macroscale model. The paper by Long and Phoonsiri(Chem.Eng. Comm.) shows the preliminary response of the model for the macroscale. The transient reponse of the model is featured.

(295e) Optimization Driven Identification of Genetic Perturbations Accelerating the Convergence of Model Parameters in Ensemble Modeling of Metabolic Networks

Ali R. Zomorrodi[1], Jimmy G Lafontaine Rivera[2], Thomas Wasylenko[3], Ali Khodayari[4], Gregory Stephanopoulos[3], James C. Liao[5] and Costas D. Maranas[6], (1)Chemical Engineering, The Pennsylvania State Univeristy, University Park, PA, (2)Chemical and Biomolecular Engineering, University of California-Los Angeles, CA, (3)Chemical Engineering, Massachusetts Institute of Technology, Cambridge, MA, (4)Chemical Engineering, The Pennsylvania State University, (5)Chemical and Biomolecular Engineering, University of California Los Angeles, Los Angeles, CA, (6)Department of Chemical Engineering, The Pennsylvania State University, University Park, PA

The recently introduced ensemble modeling (EM) approach has been proven successful in bypassing the need for experimental determination of individual kinetic parameters and dynamic modeling of metabolic networks. The essence of the EM procedure involves screening the models in the ensemble by virtue of simulating genetic perturbations (i.e., knock out/down/up) resulting in a new steady state flux distribution compared to the wild-type strain. A key element affecting the success of the EM procedure is thus the ability to converge to a set of parameters that best describe the observed phenotype of the system. However, as the set of possible kinetic parameters becomes smaller following some initial screening, subsequent genetic perturbations must become more targeted. In this study we propose an optimization-based algorithm for pro-active identification of genetic/enzyme perturbations that maximally reduce the number of retained models in the ensemble after each round of model screening. The key premise here is to design perturbations that will maximally scatter the predicted steady-state fluxes over the ensemble parameterizations thus allowing for maximal elimination of models from the ensemble and faster convergence to the correct one. We demonstrate the screening ability of this procedure using a metabolic model of the central metabolism of *E. coli* and by successively identifying single, double, triple and higher order enzyme perturbations (i.e., knockouts, overexpressions or combinations thereof) that would ensure the maximum degree of flux separation between the models in the ensemble. Overall, our study provides an optimal way of designing genetic perturbations for populating the ensemble of models with relevant model parameterizations.

(295f) Mathematical Modeling of Synaptic Vesicle Trafficking in C. Elegans

Adriana San Miguel Delgadillo[1], Celine Maeder[2], Emily Wu[2], Kang Shen[2] and Hang Lu[3], (1)Chemical & Biomolecular Engineering, Georgia Institute of Technology, Atlanta, GA, (2)Stanford University, Stanford, CA, (3)Chemical and Biomolecular Engineering, Georgia Institute of Technology, Atlanta, GA

C. elegans, a soil dwelling nematode, is a highly studied multi-cellular organism that offers several experimental advantages, including a short life span, ease of culture and transparency, to name a few. Additionally, its full genome and neuronal wiring diagram are known. This, combined with current molecular biological techniques, make it an excellent model system for biological study, especially for neuroscience. Synaptic function, assembly and maintenance is one of the areas where *C. elegans* has become an ideal model system.

Synapse assembly and maintenance is determined in part by the transport of synaptic vesicles along microtubules that run along the axon and dendrite of neurons. Pre-synaptic regions are well-defined regions in the axon, where these vesicles preferentially dock onto active zones where synapses are created onto other neurons' post-synaptic regions in the dendrite or muscles at the neuromuscular junction. Many players in the localization of such pre-synaptic regions have been identified, although the regulation of vesicle trafficking and its influence on the final location of active zones is far from understood. We generated a simple mathematical model to describe experimental data obtained for the dynamic transport of synaptic vesicle precursors in *C. elegans* obtained with neuron-wide time-lapse microscopy. The model describes synaptic vesicle transport in different regions of the neuron from a mass balance approach, and predicts that new, previously unexplored parameters are relevant for vesicle transport and distribution. Additionally, sensitivity analysis of the model hints at the most relevant parameters that ultimately determine transport and the final, steady state concentration of synaptic vesicles, as well as synapse location.

(295g) High-Content Movement Analysis As a Diagnostic Tool in C. Elegans

Peter B. Winter, Chemical & Biological Engineering, Northwestern University, Evanston, IL

Many neurodegenerative diseases manifest themselves through a loss of motor control and give us information about the underlying disease. This loss of coordination is observed in humans and in the model organisms used to study neurodegeneration. In *Caenorhabditis elegans,* there is an extensive genetic library of strains that lack functional neuronal signaling pathways and expressing proteins associated with neurodegenerative diseases. While most of these strains have decrease motility or cause paralysis, relatively few have been screened to look for more subtle changes in motor control such as stiffness, twitching, or other changes in behavior. I use high-resolution position and posture data to automatically analyze the movement of worms from different genetic backgrounds and characterize 14 movement characteristics. By creating a quantitative mapping between the movement characterization and an online database of gene annotation, gene expression, and anatomy, I aim to predict a likely set of cellular and molecular disruptions. This work provides a proof of concept for the use of detailed movement analysis to uncover novel disruptions in certain motor control processes. Knowledge of the molecular origin of these disruptions provided by our understanding of *C. elegans* genetics and physiology could lead to new diagnostic and therapeutic targets for neurodegenerative disease.

(295h) Insilico Multiscale Models for Identifying Driver Versus Passenger Mutations in Cancer Progression

Ravi Radhakrishnan, Department of Bioengineering, University of Pennsylvania, Philadelphia, PA

We describe current advances in multiscale computational modeling and simulation approaches in our laboratory combining models in structural biology and those in systems biology. The overall aim is to build quantitative models of signaling networks while retaining the crucial elements of molecular specificity, which appear to be highly relevant to profiling and predictive modeling of clinical mutations in many transformed cell lines. We discuss current and emerging experimental and computational methods, particularly focusing on hybrid and multiscale methods, which we have developed, and highlight several applications in cell signaling with quantitative and predictive capabilities. The scope of such models range from delineating protein-protein interactions to describing clinical implications in the context of mutations in ErbB family receptors (EGFR, HER2, HER3, and HER4) in non-small-cell lung cancer, and ALK receptor in pediatric neuroblastoma.

Related References:

1. Molecular Systems Biology of ErbB1 Signaling: Bridging the Gap through Multiscale Modeling and High-Performance Computing, Andrew Shih, Jeremy Purvis, R. Radhakrishnan, Molecular Biosystems (A Royal Society of Chemistry Journal), 4: 1151-1159, 2008. Pubmed ID: 19396377

2. Molecular Dynamics Analysis of Conserved Hydrophobic and Hydrophilic Bond Interaction Networks in ErbB Family Kinases, A. Shih, S. E. Telesco, S. H. Choi, M. A. Lemmon, R. Radhakrishnan, Biochemical Journal, 2011, **436**(2), 241-251. Pubmed ID: 21426301.

3. Analysis of Somatic Mutations in Cancer: Molecular Mechanisms of Activation in the ErbB family of Receptor Tyrosine Kinases, A. J. Shih, S. E. Telesco, R. Radhakrishnan, Cancers, 2011, 3(1), 1195-1231; doi:10.3390/cancers3011195. Pubmed ID: 21701703.

4. A Multiscale Modeling Approach to Investigate Molecular Mechanisms of Pseudokinase Activation and Drug Resistance in the HER3/ErbB3 Receptor Tyrosine Kinase Signaling Network, S. E. Telesco, A. J. Shih, F. Jia, R. Radhakrishnan, Molecular Biosystems (RSC Journal), 2011, 7 (6), 2066 - 2080. DOI: 10.1039/c0mb00345j. Pubmed ID: 21509365.

(322a) Magnetic Separation of Algal for Biofuel Production

Jeffrey J. Chalmers, William G. Lowrie Department of Chemical and Biomolecular Engineering, The Ohio State University, Columbus, OH, Wei Xue, Chemical and Biomedical Engineering, The Ohio State University, Columbus, Jie Xu, Chemical and Biomolecular Engineering, The Ohio State University, Columbus, OH, Maciej Zborowski, Department of Biomedical Engineering, Cleveland Clinic, Cleveland, OH and Brad Postier, Phycal, St. Louis, MO

Phycal is an algal biofuels company focusing on a hybrid photoautotrophic/heterotrophic open pond system using freshwater algal species. To **potentially reduce costs associated with algal biofuels production and/or improve the quality of intermediate steps through improved efficiencies downstream, Phycal is d**eveloping strains that can transport and store significantly more iron than wild-type strains. The potential reward is that they can out-compete contaminating species (*e.g*., algae, bacteria, protozoa, and *etc*.) for the available iron. Once stored, the iron is in the form of magnetite inside ferritin complexes that are paramagnetic. Thus, in addition to the advantage of iron scavenging, these strains can be manipulated magnetically with the potential of magnetic separation.

In this presentation we will present progress on our Phase II SBIR work to 1) characterize the magnetic susceptibility of genetically engineered and intrinsic magnetic susceptibility as a result of culture conditions, and 2) pilot scale separation systems to "dewater" algae cultures.

(322b) A Novel Taylor-Couette Photobioreactor for Energy Efficient Micro Algae Cultivation

Bo Kong, Department of Chemical and Biological Engineering, Iowa State University, Ames, IA and **R. Dennis Vigil**, Iowa State University, Ames, IA

Numerous studies have demonstrated that algal growth rate and light utilization efficiency is dramatically impacted by the cycling in the intensity of light exposure as cells move between well-illuminated photic regions near a photobioreactor external surface and the darker regions of the reactor. In particular, when the characteristic frequency of the light/dark cycles is greater than approximately 1 Hz, the "flashing light effect" can boost cell growth rates 50~100% compared to growth rates obtained without light/dark cycling. However, high frequency light/dark cycles are not easy to achieve in large-scale reactors and may be one of the reasons for low yields. In commonly used flat panel bubble column and airlift reactors, the random nature of the agitation provided by the gas bubbles produces a wide spectrum of random light/dark frequencies. In the work presented here, we make use of robust and coherent hydrodynamic structures known as Taylor vortices that arise in the annular region of a Taylor-Couette flow cell (concentric cylinders with the inner cylinder rotating) in order to rapidly and periodically shuttle algal cells between photic and non-photic regions of the reactor. Because the fluid agitation in a Taylor vortex algal reactor is not dependent on bubble-induced turbulence, much lower carbon dioxide flow rates can be used and the carbon capture efficiency is significantly higher than in conventional algal reactors. Here, we present experimental results (using Chlorella vulgaris) and analyses based upon computational fluid dynamics simulations to demonstrate that even at relatively low light-dark cycling frequencies, the Taylor vortex algal reactor offers higher biomass yield than can be achieved in corresponding bubble column experiments. Furthermore, we consider important factors for design and scale up of Taylor vortex algal reactors.

(322c) Effect of Growing Conditions On Algal Carbohydrate to Butanol Production

Alice C. Jernigan, Chemical Engineering, University of Arkansas, Fayetteville, AR and Christa N. Hestekin, Ralph E. Martin Department of Chemical Engineering, University of Arkansas, Fayetteville, AR

There are several environmentally sound reasons to use algae as a feedstock for biofuels. The fact that they are readily abundant, can absorb unwanted elements out of waste water, and can use carbon dioxide as a carbon source are just a few of the reasons that algae are being examined as one of the most likely medium to replace petroleum fuels. An issue of some concern is that microalgae have a high rate of growth in the summer and lower rates in the winter, which can cause a depletion of feedstock for a bioreactor. We are using a species of *Clostridium* bacteria to convert algal sugars to butanol in a bioreactor. In order to facilitate localized bioreactors that can operate year-round, we examined dried and stored stocks of algae taken during a six month harvesting period and examined the effects of climate, location, and processing on the amount of carbohydrates extracted from the algae and converted into butanol.

(322d) Photoautotrophic Growth and Lipid Production Kinetics of the Microalgae Scenedesmus Dimorphus

Joanne Belovich, Jacob Schwenk, Christopher Hardulak and John Van Blargan, Cleveland State University, Cleveland, OH

The design and optimization of culture systems for microalgae for biofuel production are complicated by the conflicting requirements for high biomass productivity and high oil content. Reactor models that are needed for this optimization rely on models of the algae growth and lipid productivity kinetics. Unlike heterotrophic growth of most microorganisms, photoautotrophic growth of algae can have multiple limiting nutrients, further complicated by potential light limitation. Factorial experimental designs were conducted to identify the nutrients, and their concentration ranges, that are potentially limiting for growth and oil production of the microalgae *Scendesmus dimorphus*. The individual and interactive effects of six groupings of the components of 3N-BB medium on growth rate, lipid content, and total biomass yield of *S. dimorphus* were determined. Trace metal and vitamin concentrations were reduced to 1/6 the level of 3N-BB medium without adversely affecting growth rates and biomass concentration, while concentrations of $CaCl_2$ and K_2HPO_4/KH_2PO_4 were reduced to 1/10 that of 3N-BB without adversely affecting biomass and lipid concentrations. Lipid productivity was maximized at the lowest $NaNO_3$ concentration (1/10 that in 3N-BB) and independent of $MgSO_4$ concentration, while the interaction of these two chemicals enhanced biomass concentration. Initial substrate experiments were then conducted to determine in greater detail the effects of magnesium sulfate, sodium nitrate, and potassium phosphate on biomass growth. Kinetic models for growth rate and substrate consumption were developed from the data and validated. These fundamental kinetic models can then be used in a variety of reactor styles and configurations in order to identify the system and operating parameters

(322e) Use of Sodium Bicarbonate for Efficient Carbon and Water Management for Autotrophic Microalgae Cultivation in Open Pond System

Jinsoo Kim, School of Energy, Environmental, Biological and Medical Engineering, University of Cincinnati, Cincinnati, OH, Joo-Youp Lee, School of Energy, Environmental, Biological, and Medical Engineering, University of Cincinnati, Cincinnati, OH and Ting Lu, Wastewater Treatment Division, Metropolitan Sewer District of Greater Cincinnati, Cincinnati, OH

Dissolved inorganic carbon including carbon dioxide(aq) and bicarbonate ion (HCO_3^-) is an essential carbon source for the autotrophic growth of microalgae. However, the use of CO_2 gas only cannot provide a sufficient amount of dissolved inorganic carbon concentrations in a narrow pH window suitable for the growth. In this study, sodium bicarbonate was used to increase dissolved inorganic carbon concentrations in conjunction with CO_2 gas. Added CO_2 gas was able to suppress the increased pH values as a result of bicarbonate ion uptake and to supply additional dissolved inorganic carbon. The fates of inorganic carbon and sodium ion were studied by taking mass balances for the two components. The effects of sodium ion on CO_2 gas absorption and algal productivity will be presented for sustainable water reuse and autotrophic microalgal culture in a typical open pond system. A preliminary economic analysis of the use of sodium bicarbonate based on the experimental results will also be presented.

(322f) Transgenic Expression of a Bacterial Exo-Acting Intracellular α-Amylase in the Chlamydomonas Reinhardtii Chloroplast

Xiaoqing Wang[1], Barbara Sears[2], Yan Susie) Liu[3] and Wei Liao[1], (1)Biosystems and Agricultural Engineering, Michigan State University, East Lansing, MI, (2)Plant Biology, Michigan State University, East Lansing, MI, (3)Department of Biosystems and Agricultural Engineering, Michigan State University, East Lansing, MI

Microalgae with their relatively high starch and lipid contents are one of the most promising non-food-crop-based feedstocks for biofuels production. To date, most studies have focused on accumulating algal lipids for biodiesel production and some have focused on hydrogen generation. Very little attention has been paid to simultaneously converting algal carbohydrate and lipids into biofuels. Although some algal species are able to accumulate significant amounts of carbohydrates (mainly starch) and lipid, their rigid cell wall retards the extraction of those carbon-rich compounds. In order to overcome this barrier, this study constructed an amylase-producing strain of the unicellular green alga *Chlamydomonas reinhardtii*, with the goal of converting algal starch into soluble sugars *in vivo*, which would facilitate their recovery. An exo-acting intracellular thermophilic α-amylase AmyB from the hyperthermophilic bacterium *Thermotoga neapolitana* was used as the amylase gene of choice. The AmyB gene was synthesized with codon usage optimized for the *C. reinhardtii* chloroplast; it was cloned into chloroplast expression vector p666 to yield plasmid p666AmyB. The vector p666AmyB was inserted into the site between the *psaB* and *atpA* genes on the chloroplast genome of *C. reinhardtii* by micro-particle bombardment, without interruption of chloroplast gene function. PCR analysis indicated that *AmyB* was integrated into the chloroplast genome of the transformants, and western blotting demonstrated the AmyB protein accumulation. The activity of AmyB can reach 17.4 U/g dry biomass. Thus, α-amylase was successfully expressed within the chloroplast, accomplishing an important step towards improving the ability to harness and harvest the carbon-compounds synthesized by micoalgae.

(352a) Modeling the Molecular Crosstalk Between a Growth Factor and a Differentiation Inducing Receptor

Konstantinos Biliouris, Chemical Engineering and Materials Science, University of Minnesota, Minneapolis, MN, Harinder Singh, Department of Discovery Immunology, Genentech Inc., South San Francisco, CA and Yiannis Kaznessis, Department of Chemical Engineering and Materials Science, University of Minnesota, Minneapolis, MN

The Interleukin-7 receptor (IL-7R) and the precursor to the B cell antigen receptor (pre-BCR) function to regulate the balance between pre-B cell proliferation and their differentiation. The former signaling system promotes cell survival and proliferation whereas the latter stimulates antibody light chain gene rearrangement thereby enabling the generation of mature B-lymphocytes. The architecture of the gene regulatory network (GRN) orchestrating the interplay between the two signaling systems has been recently elucidated [1].

IL-7 signaling is coupled to STAT5, PI3K and Akt to promote cell proliferation. On the other hand, pre-BCR signaling is coupled to Syk, BLNK, FoxO1 and Pax5, which induce cellular differentiation. Weakening of IL-7R signaling, or an alteration in pre-BCR signaling, results in cell-cycle arrest and cell differentiation [1].

The complexity level of this GRN is high. Therefore, guided by experimental observations, we are developing a mathematical model to fully grasp the multifaceted network underlying the crosstalk between IL-7R and pre-BCR. It appears that this GRN shares a key characteristic with previously studied systems [2,3]. This characteristic refers to the mutual (and sometimes indirect) repression between two primary modules, FoxO1-Pax5 and IL-7R, and gives rise to bistable behavior when described by non-linear dynamics. Interestingly, preliminary results indicate that the architecture of the presented GRN does generate bistability under certain parameters. The two stable steady states, namely cell proliferation and differentiation, are separated by unstable steady states. As discussed previously [3], these unstable steady states cannot be observed experimentally and can only be explored mathematically.

Our mathematical framework investigates all the possible states of the IL-7/pre-BCR GRN. Sensitivity analysis with respect to IL-7 concentration allows for defining its threshold for flipping this molecular circuit (from cell differentiation to proliferation and vise versa) and the parameters that govern it. Our computational predictions will be validated against experimental measurements aiming to thoroughly explain the dynamics dictating whether cells undergo differentiation or proliferation.

1) K. Ochiai, M. Maienschein-Cline, M. Mandal, J.R. Triggs, E. Bertolino, R. Sciammas, A.R. Dinner, M.R. Clark, and H. Singh, Nature Immunology, 13(3):300-307, 2012.

2) P. Laslo, C.J. Spooner, A. Warmash, D.W. Lancki, H.J. Lee, R. Sciammas, B.N. Gantner, A.R. Dinner,and H. Singh, Cell, 126(4):755-766, 2006.

3) D. Muzzey and A. van Oudenaarden, Cell, 126(4):650-652, 2006.

(352b) Coordination of Quorum Sensing with Cell Motion

David N. Quan, College Park, Institute for Biotechnology and Bioscience Research, College Park, MD; Bioengineering, University of Maryland at College Park, College Park, MD and William E. Bentley, Fischell Department of Bioengineering, University of Maryland, College Park, College Park, MD

Coordination of Quorum Sensing and Cell Motion

David N. Quan, William E. Bentley

Through computer aided modeling, emergent phenomenon may be elucidated—how the interaction of a billion simple answers (forged through reductionist approaches) create a cohesive cellular/organismal response can be understood. A first step, of course, is to investigate the isolated interaction of two simple answers. Here, we investigate phenomenon arising from the overlap of quorum sensing (QS) and cell motion.

QS is a concentration dependent phenomenon used by bacteria to sense local diffusion conditions or local cell concentration. Among the few known and first identified QS agents are the autoinducers (AI) AI-1 and AI-2. AI-1 is an umbrella term for species specific AHL's secreted from *Vibrio* bacteria which, after achieving a sufficient concentration, trigger a positive feedback mechanism. AI-2 is much more broadly distributed QS agent working either through LuxPQ, a two component regulatory system, or the more elaborate LuxS-regulated (Lsr) system that recompartmentalizes AI-2 to achieve quorum activation. Because these QS agents are small molecules, the effective concentration that the cells are exposed to is highly dependent upon local convection and diffusion dynamics, which are influenced—especially stochastically—by cell transport (motility). Here, we predict nontrivial phenomenon arising from the confluence of these transports (QS agent and cell) based on *in silico* experimental results.

Cells were treated as autonomous single agents that either swam or glided within a finite element environment through which AI-2 diffused. At each time point, effective QS agent concentration was integrated across local grids, QS state (whether the genes were switched on) was evaluated, QS agent exchange between the cell and the media took place, cells divided and moved stochastically, and QS agent diffused between grids.

Firstly, we successfully recapitulated the supernova-like activity pattern of a surface attached growing colony under the influence of AI-1 QS. Next, we used an ODE model of AI-2 activity to inform the stochastics governing Lsr behavior. Applying the forms of the ODEs to colony growth in our finite element/single agent scheme resulted in a stable, abrupt checkered activation. We extended this particular model by adding AI-2 synthase (LuxS) controlled by the *lsr* promoter, and predict that the proportion of QS+ cells is enhanced depending on the plasmid copy number and where in parameter space the system operates. Lastly, we modeled the overlap of Lsr based QS dynamics with swimming cells that chemotax according to AI-2 gradient. In these simulations, AI-2 and cell transport were limited by virtual boundaries. Over the course of the experiment, cells/AI-2 coalesced in random clusters along these boundaries, especially in the corners, trapped by a positive feedback loop. This behavior persisted until QS activation, at which point the driver of chemotaxis was swept up and bacteria within these clusters redispersed.

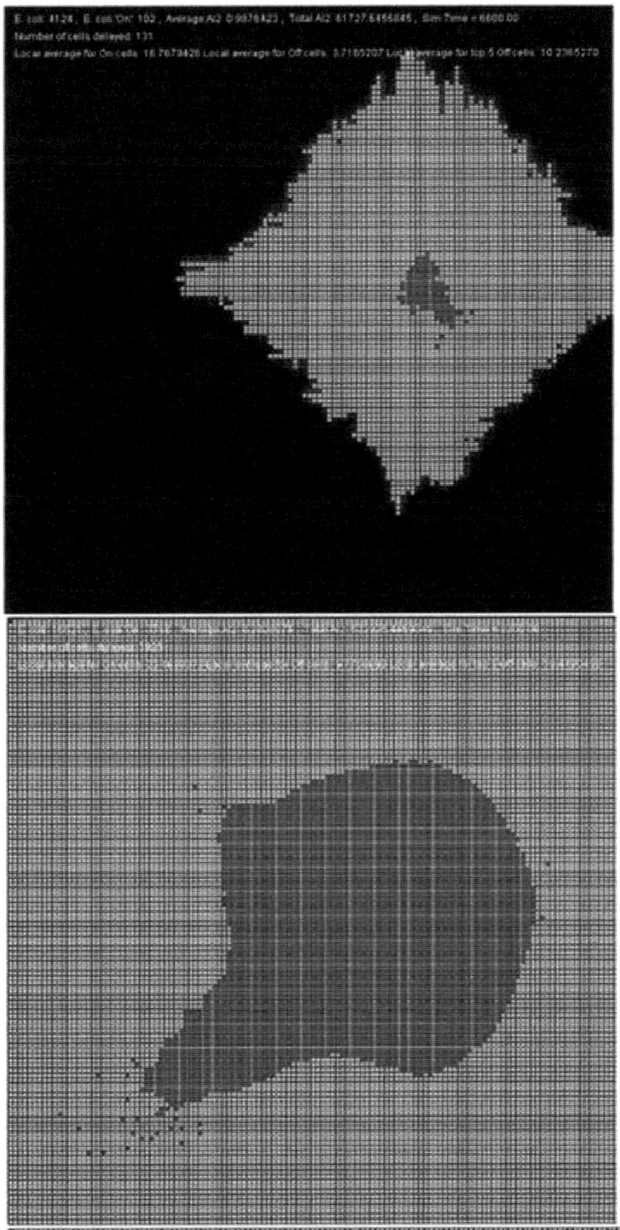

Figure 1a and b. This simulation modeled AI-1 supernova like expansion of QS expression. Cells are colored yellow-orange, while QS "On" cells are colored red. In the first figure, one sees the incipient outward expansion of expression at the center of the cell cluster,. At the later time point this expansion has continued apace.

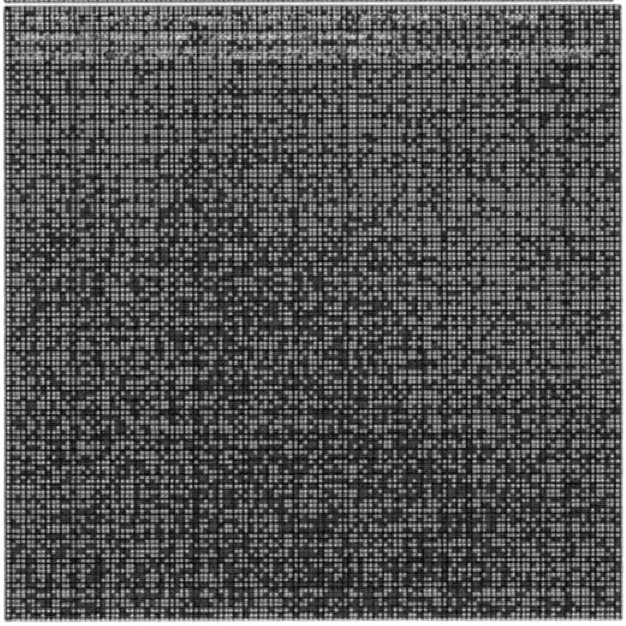

Figure 2. This simulation modeled AI-2 checkered activation in Lsr-based QS. Cells are colored yellow-orange, while QS "On" cells are colored red. Density of activation is highest where the cell colony began its growth from. The pattern, once developed, is stable over time.

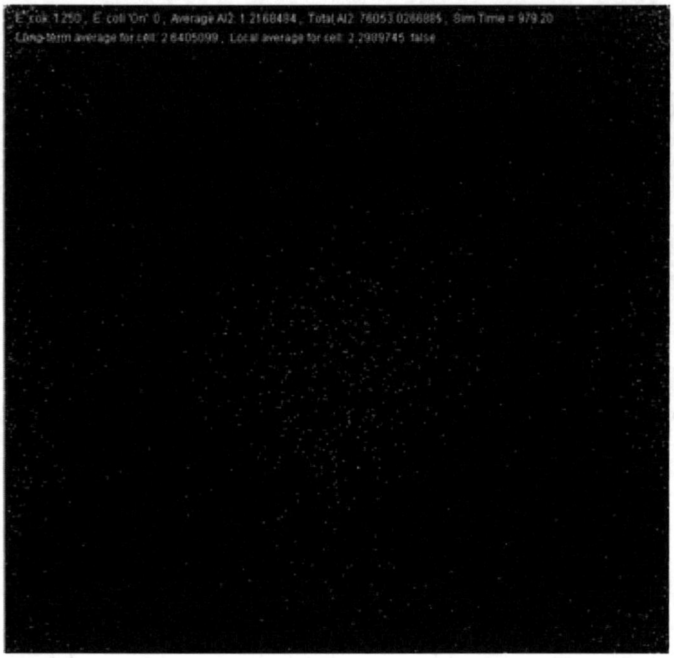

Figure 3. This simulation models bacteria free swimming in response to AI-2 chemotaxis. Cells are colored yellow-orange. Cells are concentrated at the center, where a bolus was originally deposited, and at the edges/corners. Once Lsr-based QS turns on, the cells redisperse evenly in the environment.

The largest and most comprehensive mouse metabolic reconstruction to date contains 1,415 metabolic genes accounting for reactions in eight cellular compartments (cytosol, mitochondrial, extracellular, golgi, lysosome, ribosome and nucleus). Despite these capabilities, the current model lacks the ability to simulate several essential metabolic functions. Missing links in this mouse metabolic network model were identified to ensure fundamential functionality of generic cell metabolism. Tissue-specific models were built with the improved generic model. There are three tissues that we are interested in for this project: the liver (the major metabolic cell type is hepatocyte), the skeletal muscle (consists mainly of myocytes) and the adipose tissue (consists primarily of adipocytes). These three tissues are the major subsystems glucose circulates through after it enters the blood. The metabolic tissue models were constructed specifically for the widely studied C57BL/6J mouse strain using an automatic tissue reconstruction algorithm. We will then identify reactions specific to each tissue of interest and simulate these functions in tissue-specific models. These models will then possess these metabolic abilities in addition to the universal tests for the generic model. The last part of this project involves introducing a blood compartment into the multi-tissue system to connect different tissues and model tissue-tissue interactions.

(352d) Development of a Metabolite Stress-Response Model in Solventogenic Clostridia by Coupling Multiple –Omic Data with a Genome-Scale Model

Eleftherios T. Papoutsakis[1], Shawn W. Jones[2], Keerthi P. Venkataramanan[2], Blake C. Meyers[3], Sridhara G. Kunjeti[2], Kevin P. McCormick[2], Kelvin H. Lee[1], Shuyu Hou[2], Maciek R. Antoniewicz[4], Jungik Choi[2], Jennifer Au[2], Costas D. Maranas[5], Patrick F. Suthers[5], Qinghua Wang[2], Hongzhan Huang[6] and Cathy H. Wu[6,7], (1)Department of Chemical and Biomolecular Engineering & Delaware Biotechnology Institute, University of Delaware, Newark, DE, (2)University of Delaware, Newark, DE, (3)Department of Plant and Soil Sciences & Delaware Biotechnology Institute, University of Delaware, Newark, DE, (4)Department of Chemical and Biomolecular Engineering, University of Delaware, Newark, DE, (5)Department of Chemical Engineering, The Pennsylvania State University, University Park, PA, (6)Delaware Biotechnology Institute, University of Delaware, Newark, DE, (7)Department of Biochemistry and Molecular & Cellular Biology, Georgetown University Medical Center, Washington, DC

The toxic-metabolite stress response is a problem of major and general importance in all microbial systems of interest to bioenergy production. Among the potential microbial hosts for biofuel production, solventogenic clostridia are of major importance because of their inherent ability to utilize a large variety of substrates (hexoses, pentoses, oligosaccharides, xylan, and starches) with minimal to no catabolite repression. Here, we focus on understanding and modeling the stress response of *Clostridium acetobutylicum* to two important toxic metabolites: butanol and butyrate. Butanol is the major metabolite produced during stationary phase, while butyrate is primarily produced during exponential growth.

In order to model the stress response to these metabolites, *C. acetobutylicum* cultures were grown to mid-exponential phase and then stressed with four levels of the metabolite of interest (no, low, medium, and high levels of stress). Samples were collected following the stress to look at the cellular response to the different stress levels. The transcriptional response was measured using both genomic microarrays and deep sequencing (RNA-seq) using the Illumina technology. The goal of using these complimentary technologies is to not only assess gene expression patterns affected by the metabolite stress, but also examine and model changes at the molecular level impacted by stress, such as changes in expressed operon mRNAs, transcriptional start cites, as well as stress-specific RNA degradation. Furthermore, proteomic changes were measured using the iTRAQ technology to quantitatively evaluate protein expression changes, especially because the key mechanism of regulation of the universal stress response system (the HSP system) is largely regulated at the protein level by protein-protein interactions. Finally, guided by the transcriptional data, metabolic flux changes of core programs within the cells were determined using ^{13}C-labeled chemicals. The goal of these measurements is to quantitate how cells re-allocate its energy and carbon resources under stress, whereby a large number of programs are downregulated but a few are upregulated. These data were then integrated within a second-generation genome-scale model, which incorporates algorithms to regulate the model based on gene and protein expression, protein-protein interactions, and metabolic flux regulation. We ultimately aim to use this model to design a semi-synthetic stress response system that enables the cells to produce toxic metabolites at higher rates and titers.

(352e) Metabolic Engineering in Silico Enabled by Manipulating Metabolic Pathway Flux Ratios

Ryan S. Senger, Biological Systems Engineering Department, Virginia Tech, Blacksburg, VA

A new method of constraining genome-scale metabolic flux models called "flux balance analysis with flux ratios" (FBrAtio) has enabled significant progress in metabolic engineering *in silico*. Despite their extraordinary promise and popularity, it is well established that genome-scale models suffer from having a large number of degrees of freedom. Several tools have been developed to address reaction directionality and gene regulation in order to reduce the number of solutions to these models. However, the large solution space associated with these models has thus far prevented widespread use of genome-scale models in a "predictive" capacity that is required of metabolic engineering *in silico*. The novel FBrAtio approach considers a metabolite of a metabolic network as a "node." At a metabolic branch point, multiple competing enzymes can consume a metabolite. Ultimately, thermodynamics determine how this metabolite is distributed among competing enzymes, and genetic regulation (with spatial partitioning) determines the concentrations of enzymes present. This distribution of a metabolite among competing enzymes is easily specified by the flux ratio concept. Thus, the flux ratios at key branch points in a metabolic network can be optimized for over-production of a desired product and lead to "predictive" modeling using genome-scale metabolic flux models. This is significant because flux ratios can be specified experimentally through regulatory engineering. The FBrAtio algorithm returns a "fine-tuned" metabolic engineering strategy that involves targeted levels of gene over-expression and knock-down in addition to knockout. The flux ratios approach returns a quantitative requirement of up- or down-regulation of targeted enzymes at key metabolic branch points. In this presentation, the concepts of flux ratios are explained in detail and are applied to model systems. Experimental validations of FBrAtio designed metabolic engineering strategies are presented for (i) the over-accumulation of cellulose by Arabidopsis, and (ii) over-production of bio-butanol and ethanol by solventogenic clostridia. In addition, the application of FBrAtio is also presented for the case of optimizing flux through synthetic pathways installed in cyanobacteria.

(352f) Host Metabolic Interaction During Viral Infection: A Model of Bacteriophage T7 In E. Coli

Elsa Birch, Chemical Engineering, Stanford University, Stanford, CA, Nicholas Ruggero, Stanford University, Stanford, CA and Markus Covert, Bioengineering, Stanford University, Stanford, CA

Bacteriophage progeny, and in fact all viruses, are metabolic products of the infected host cell. Viral replication competes with the host for use of both synthesis machinery and metabolic precursors. A thorough understanding of the host-viral metabolic interaction will add significantly to the basic understanding of phage, as well as facilitate efforts to restrict or employ their spread. *E. coli* metabolism has been modeled extensively using Flux Balance Analysis, and phage T7 replication has been described in structured detail using ordinary differential equations. Developing an integrated simulation method requires additions to the independent component models, as well as construction of a new algorithm to enforce mutual limitation by the two models. Our integrated simulation enables exploration of the constraints placed on phage replication by the host metabolic capacity, as well as the dynamic perturbation of the host network state. The model predicts nutritional perturbations that have a strong impact on viral replication, which we demonstrate experimentally.

(352g) In Silico Model of Suppression and Desynchronization of Peripheral Clock Genes in Human Endotoxemia

Panteleimon D. Mavroudis, Chemical and Biochemical Engineering, Rutgers University, Piscataway, NJ, Steve E. Calvano, Department of Surgery , UMDNJ-Robert Wood Johnson Medical School, New Brunswick, NJ and Ioannis P. Androulakis, Department of Chemical and Biochemical Engineering, Rutgers, The State University of New Jersey, Piscataway, NJ

Circadian rhythmicity is maintained in the human body through a hierarchical web of cell autonomous and self-sustained molecular clocks that drive transcriptional oscillations to a group of genes commonly known as clock genes. These biological clocks are synchronized with environmental light/dark cycles and provide circadian information to numerous physiological processes including sleep/wake activity, feeding rhythms, and body temperature in order to optimize organism's performance (Dibner, Schibler et al. 2010). The immune system is also under circadian regulation since a wide range of immune parameters such as the level of red blood cells, peripheral blood mononuclear cells and cytokines undergo daily fluctuations. Data further suggest that the immune system can also provide inputs to clock genes network that can adjust its activity and ultimately modulate numerous physiological rhythms such as circadian production of cytokines and hormones. Importantly, acute as well as chronic stress conditions induce significant alterations in the rhythmic expression of peripheral clock genes disturbing their synchronized dynamics further uncoupling them from light mediated body rhythm (Haimovich, Calvano et al. 2010).This loss of body's synchronization by environmental cues, may negatively impact health by impairing host defense against viral infection and other diseases. To further assess these effects, in this work we discuss a mathematical model of human endotoxemia where we evaluate the responses of peripheral clock genes in response to endotoxin stimulus mostly relative to their intercellular synchronization, as well as the rhythmic implications to critical immune compartments such as cortisol's and pro-/anti-inflammatory cytokines circadian rhythm.

Elective administration of endotoxin (lipopolysaccharides, LPS) to otherwise healthy human volunteers has been used to study systemic inflammation and gain insight into behaviour of inflammatory mediators encountered in acute, as well as chronic, inflammatory diseases. Human endotoxemia precipitates signs and symptoms similar to what is observed clinically in sepsis patients and early trauma patients. While we do not argue that the human endotoxin challenge model precisely replicates an acute infectious or septic condition, we believe that endotoxemia does serve as a useful model of TLR4 agonist-induced systemic inflammation by providing a reproducible experimental platform tying systemic inflammation to physiological signal generation (Haimovich, Reddell et al. 2010). Based on our prior work on modeling human endotoxemia (Scheff, Mavroudis et al. 2011), we aim to develop a multi-compartment mathematical model that incorporates the hypothalamic pituitary adrenal (HPA) gland's mediated circadian secretion of cortisol, and the rhythmic secretion of pro- (P) and anti- (A) inflammatory cytokines that is mediated by the peripheral clock gene (PCG) network of immune cells. Administration of LPS leads to the activation of the NF-kB signaling pathway that ultimately leads to the upregulation of the pro-inflammatory cytokines (Scheff, Calvano et al. 2010). Increase pro-inflammatory cytokines further mediates the inflammatory signal to HPA axis that regulates the circadian expression of cortisol and ultimately the expression of anti-inflammatory cytokines that serve as the immunoregulatory component of the system in order to restore homeostasis. Recently, we showed that cortisol circadian rhythm apart from playing a key anti-inflammatory role in the inflammatory response, it synchronizes the expression of a population of peripheral clock genes in an amplitude and frequency dependent manner (Mavroudis, Scheff et al. 2012). In addition, recent experimental data indicate that pro-inflammatory cytokines such as TNF-α and IL-1 which rhythmic production is mediated by PCGs (Keller, Mazuch et al. 2009), downregulate the expression of E-box mediated transcription of PCGs (Cavadini, Petrzilka et al. 2007) forming in such a way a feedback regulatory mechanism between clock genes and pro-inflammatory cytokines.

For this paper, we aim to investigate the bidirectional link between P and PCGs. We hypothesize that pro-inflammatory cytokines inhibit the E-box mediated transcription of *Per* and *Cry* clock genes by blocking CLOCK/BMAL1 transcription factor binding to this region. The circadian secretion of P is mediated by the ensemble rhythm of a population of PCGs that are entrained by F. The endotoxin stimulus in our model is hypothesized to drive the upregulation of P that further induces F that apart from entraining PCGs, is assumed to regulate the A compartment of our model that ultimately leads to inhibition of the pro-inflammatory response by inhibiting P. Our model indicates that administration of LPS ultimately leads to a loss of intercellular phase coherency to the population of PCGs. This intercellular desynchronization is time of day dependent, pointing higher values at these times of day that PCGs reach their zenith levels.

Our work represents a step towards understanding the biological mechanisms of human endotoxemia. By evaluating the underlying dynamics resulting from the link between P and PCGs we start deciphering the mechanistic details leading to body's loss of rhythmicity on stress conditions. The connections between circadian clock and immune system of the body is of high importance with strong current interest and our work contributes to this field with new interesting facts.

References:

Cavadini, G., S. Petrzilka, et al. (2007). "TNF-Î± suppresses the expression of clock genes by interfering with E-box-mediated transcription." Proceedings of the National Academy of Sciences **104**(31): 12843-12848.

Dibner, C., U. Schibler, et al. (2010). "The mammalian circadian timing system: organization and coordination of central and peripheral clocks." Annu Rev Physiol **72**: 517-49.

Haimovich, B., J. Calvano, et al. (2010). "In vivo endotoxin synchronizes and suppresses clock gene expression in human peripheral blood leukocytes *." Critical Care Medicine **38**(3): 751-758 10.1097/CCM.0b013e3181cd131c.

Haimovich, B., M. T. Reddell, et al. (2010). "A novel model of common Toll-like receptor 4- and injury-induced transcriptional themes in human leukocytes." Crit Care **14**(5): R177.

Keller, M., J. Mazuch, et al. (2009). "A circadian clock in macrophages controls inflammatory immune responses." Proceedings of the National Academy of Sciences **106**(50): 21407-21412.

Mavroudis, P. D., J. D. Scheff, et al. (2012). "Entrainment of peripheral clock genes by cortisol." Physiol Genomics.

Scheff, J. D., S. E. Calvano, et al. (2010). "Modeling the influence of circadian rhythms on the acute inflammatory response." J Theor Biol **264**(3): 1068-76.

Scheff, J. D., P. D. Mavroudis, et al. (2011). "Modeling autonomic regulation of cardiac function and heart rate variability in human endotoxemia." Physiol Genomics **43**(16): 951-64.

(352h) Coarse-Grained Modeling of Molecular Release from an Actin Network

John Kang, Computational Biology, Carnegie Mellon University, Pittsburgh, PA; School of Medicine, University of Pittsburgh, Pittsburgh, PA, Kathy M. Puskar, Mechanical Engineering, Point Park University , Pittsburgh, PA, Russell S. Schwartz, Biological Sciences, Carnegie Mellon University, Pittsburgh, PA and Philip R. LeDuc, Mechanical Engineering, Carnegie Mellon University, Pittsburgh, PA

Recent literature provides strong evidence for a causative link between the mechanical stretching of cytoskeleton and the release of signaling molecules [1]. Understanding the link between the mechanical input, the corresponding morphological changes in the actin cytoskeleton, and the final signaling molecule release is a poorly understood yet very significant problem in the field of mechanotransduction. Here, we present a coarse-grained actin network model that simulates architectural changes in the network in response to external mechanical stimulus to understand the interplay between actin network mechanics and resulting biochemical signaling. We develop a coarse-grained actin network model and examine the intrinsic angle geometry changes under mechanical stimulation. We have previously examined actin network mechanics with a derivative of this model [2]. The network is created with a circular solution space that is considered fixed to an underlying substrate at uniformly placed perimeter nodes representing focal adhesions. Filaments are formed by crosslinking opposite focal adhesions on the periphery. Intersections formed by crosslinking represent molecular linking of actin with affiliated attached molecules (e.g. filamin A and FilGAP) at each of the four angles created by that intersection. Stretch is simulated by moving the upper nodes while holding the bottom nodes stationary and balancing forces on the remaining nodes.

As stretch is applied to the network, the intersecting angle distribution transitions from a more peaked to a flatter distribution while still being centered at 90°. We also examined the difference in the stretched angle relative to the non-stretched angle ("delta angle"), and observe a shift in distribution as additional stretch is applied. At 1% stretch, the delta angles are small with almost no angle changes greater than 5 degrees, but at higher levels of stretch, the observed delta angle distribution is almost uniform. These histograms reflect similar results to simulations performed by Ehrlicher et al. on their experiments even though their simulations had different overall morphologies and boundary conditions [1].

These results are being incorporated into molecular release models that represent a potentially versatile platform for examining the biophysical interactions that link mechanical stimulus at the cellular level to response at the protein level and may underlie strain-induced signaling mediated by the cytoskeleton.

References

1. Ehrlicher et al. (2011) Mechanical strain in actin networks regulates FilGAP and integrin binding to filamin A. Nature 478, 260-261.

2. Kang J et al. (2011) Response of an actin filament network model under cyclic stretching through a coarse grained Monte Carlo approach. J Theor Biol 274, 109-119.

(430a) Integrating High-Throughput Data with Biochemical Networks Identifies Functional Regulatory Interactions

Sriram Chandrasekaran[1,2] and Nathan D. Price[1], (1)Institute for Systems Biology, Seattle, WA, (2)Biophysics, University of Illinois at Urbana Champaign, Urbana, IL

Emerging high-throughput technologies allow the rapid measurement of potential regulatory interactions, yet their ability to predict cellular phenotype is still poor, mainly due to the lack of methods to directly connect measured interactions to observable phenotypes, and the presence of a plethora of false-positive interactions. We present an approach – Gene Expression and Metabolism Integrated for Network Inference (GEMINI) – that effectively assesses interactions by using a biochemically-detailed metabolic network to predict the phenotypic consequences of regulation on cellular metabolism. By comparing predictions to phenomics data, GEMINI can produce a refined network model that is significantly better at recalling true interactions than conventional expression-based methods and show increased accuracy in predicting gene knockout phenotypes under different conditions. GEMINI represents an attempt to curate the inference of regulatory interactions by leveraging on a metabolic network—an approach that incorporates multiple layers of biological context and a mechanistic framework to the problem of regulation.

References

[1] Chandrasekaran *et al*, Behavior-Specific Changes in Transcriptional Modules Lead to Distinct & Predictable Neurogenomic States, *PNAS*, 2011.

[2] Chandrasekaran & Price, Probabilistic integrative modeling of genome-scale metabolic and regulatory networks in *E.coli* and *M. tuberculosis*, *PNAS*, 2010.

[3] Barret *et al*, The global transcriptional regulatory network for metabolism in E.coli exhibits few dominant states, *PNAS*, 2005.

(430b) Large-Scale Kinetic Model of Chlamydomonas Reinhardtii Metabolism Reveals Diffuse Control Over Growth Precursors

Mariajose Castellanos, Chemical, Biochemical, and Environmental Engineering, UMBC, Baltimore, MD and Jon Bollinger, Chemical, Biochemical, and Environmetal Engineering, UMBC, Baltimore, MD

Algae are used for the industrial-scale production of chemicals spanning food supplements, pharmaceuticals, and biofuels because they are robust, photosynthetic, and can be cultivated on a large scale. Similar to other industrially-used microorganisms, it is desirable to develop an *in silico* model of algae metabolism that can be used to predict the effects of environmental and genomic perturbations and stress, which could be used to rationally identify approaches for engineering more productive strains. Dynamic metabolic models are of particular interest because they can inherently describe time-based perturbations, metabolic cycles, and complex enzyme and gene interactions. Herein we detail the development and validation of the first reported large-scale kinetic model of the *Chlamydomonas reinhardtii* metabolism, comprising four organeller compartments, 369 chemical reactions, and incorporating known inhibition kinetics for highly regulated control points (e.g. Calvin, TCA). After generating an ensemble of thousands of possible models for this system, which spanned known thermodynamic and biological limits, and screening these models for nitrogen-deprivation phenotypes observed experimentally, a model was identified that reproduced known nutrient exchange trends and observed shifts in growth precursors, energy stores, and intracellular metabolite contents under nutrient stress. A control study was then executed to identify enzymes expressions that exerted great pressure over growth-related fluxes and CO2 uptake; this study revealed that control over growth is wielded by a very diffuse set of reactions, many of which lie outside of the metabolic pathways typically included metabolic studies of photosynthetic organisms. It also suggested several genetic engineering targets for improved biomass accumulation.

(430c) Comparison of Network Structures That Confer Resilience Against Genetic Perturbations in Microbial Metabolism

Chintan Joshi and Ashok Prasad, Chemical and Biological Engineering, Colorado State University, Fort Collins, CO

Due to their relatively high rates of division, most microorganisms need to be resilient against random genetic perturbations. In the metabolic network, microorganisms can survive a large number of gene deletions by altering the flux distribution to compensate for missing reactions. Understanding the design principles of the metabolic network that confer this resilience to the organism is of great interest for metabolic engineering strategies that aim to re-direct fluxes towards commercially important products like biofuels, nutraceuticals and pharmaceuticals. We used flux balance analysis (FBA) along with minimization of metabolic adjustment (MOMA) in order to predict the effects of gene knockouts on two genome-scale models of microbial metabolism under different conditions -- that of a bacterial network (*E. coli* iAF1260 – aerobic and anaerobic growth) and a cyanobacterial network (*Synechocystis sp.* PCC6803 iJN678 – autotrophic, mixotrophic, and heterotrophic growth). In autotrophic metabolism more than 52% of the mutants were lethal, while the remainder grew at nearly optimal growth rates, which was achieved by taking up higher amounts of light. We mapped genes leading to lethal knock-outs in one network on to the other and screened for non-lethality. We found that several lethal knockouts in one organism were non-lethal in the other, in other words the two organisms had different resilience properties. We identified some key subsystems of the *E. coli* metabolic network and the electron flow machinery of Synechocystis as being primarily responsible for conferring this resilience. We found that different sets of metabolites were very tightly regulated in these organisms, and genetic mutations that led to drastic perturbations in reactions involving these metabolites proved lethal for the organism. Analysis of epistasis by carrying out double gene deletions revealed that the autotrophic environment led to a large number of aggravating gene pair interactions. However, the heterotrophic environment and aerobic growth of *E. coli* led to relatively higher buffering interactions. Further, epistasis analysis was used to reveal more differences in the association between genes and reactions of metabolic network structures of the two organisms. Our analysis allows us to make predictions about; (i) why bacterial and cyanobacterial network exist the way they do, and (ii) how these organisms might benefit out of these unique gene-reaction associations. Thus photosynthetic and non-photosynthetic organisms have different resilience properties, arising in part from the differing structures of their metabolic networks, metabolite essentiality, and gene-reaction associations.

(430d) Contrasting the Metabolic Capabilities of Cyanobacterial Strains for Assessing Bio-Production Platform Selection

Rajib Saha, Alex T. Verseput, Thomas J. Mueller and Costas D. Maranas, Department of Chemical Engineering, The Pennsylvania State University, University Park, PA

Cyanobacteria are photoautotrophic prokaryotes that exhibit robust growth under diverse environmental conditions with minimal nutritional requirements. They can use solar energy to convert CO2 and other reduced carbon sources into a variety of biofuels and chemical products. The Cyanothece genus of cyanobacteria has been shown to offer very high levels of hydrogen production and nitrogen fixation. Here we reconstruct and subsequently analyze and contrast the differences in metabolism between five Cyanothece strains, namely *Cyanothece* 7424, 7425, 7822, 8801 and 8802. We compare these phylogenetically related Cyanothece strains to assess their suitability as microbial production platforms. Reconstructed genome-scale metabolic models include experimentally verified biomass compositions, fully traced photosynthesis reactions and respiratory chains as well as balanced reaction entries and GPR associations. From the reconstructed models, we find *Cyanothece* strains share almost 90% of their reactions. Common reactions are mainly involved in nitrogen fixation, central carbon metabolism, photosynthesis and respiration, amino acid biosynthetic pathways as well as bidirectional hydrogenase activity. All but *Cyanothece* (7424 and 7822) have complete alkane production pathways. Furthermore, all these strains share a non-fermentative higher-alcohol production pathway. *Cyanothece* 7425 is phylogenetically most distant among these Cyanothece strains, reflected in the 332 unique reactions not shared with the other Cyanothece strains. By focusing on the specific differences among these Cyanothece strains, our models reveal exactly which steps are the ones absent in fermentative Butanol production pathway. Specifically, *Cyanothece* 7425 has the most complete pathway missing only the conversion between 3-Hydroxybutanoyl-CoA and Crotonoyl-CoA, while *Cyanothece* 7822 lacks a step dealing with the conversion between Crotonyl-CoA and Butanoyl-CoA. Another notable difference is the presence or absence of the glyoxylate shunt in the TCA cycle. *Cyanothece* (7425, 8801 and 8802) do not have the glyoxyate shunt, whereas *Cyanothece* (7424 and 7822) appear to have all relevant enzymes. Herein, we describe results from comparing important pathways leading to biofuel or other useful product formation followed by experimental verification.

(430e) MC^3: A Tool for Model and Constraint Consistency Checking of Stoichiometric Biochemical Network Models

Ehsan Ullah, Mona Yousofshahi, Soha Hassoun and Russell Stern, Computer Science, Tufts University, Medford, MA

Stoichiometric models are traditionally constructed manually, based on earlier models, the availability of reaction data from databases, and newly available experimental data. An example is the formulation of successive models of the Escherichia coli organism that lead to improved predictive capabilities and elucidation of phenotypic behavior. Models thus have gaps (missing reactions), and there is sometimes uncertainty about reaction directions. While some methods have been advocated to ensure the correctness of some aspects of a stoichiometric model within a particular context, there is currently no standalone computational tool that ensures model and constraint consistency. Each user is thus forced to manually sanitize his or her model, and often it is an incorrect computational result that alerts the user of the incorrectness of the model. Models released in the public domain often have undocumented issues, such as dead-end metabolites or zero-flux reactions. Additionally, there is no standard documentation protocol that each model undergoes prior to public release. The need for agreed-upon standards and for computational tools will become more pressing with the rise of the size and number of genome-scale models.

We define and formalize the notion of a valid stoichiometric biochemical network model suited for steady-state analysis, and develop a computational tool, MC^3 tool, a Model and Constraint Consistency Checker, to validate stoichiometric models. The MC^3 tool consists of two distinct checking components, one based on the results of the kernel computation for S .v =0 and one based on exercising Flux Balance Analysis. We checked the consistency of some published available models using MC^3. The checked models included those of Escherichia coli (3 different sizes), adipocyte, Chinese Hamster Ovary (CHO) cell, Leishmania major, and Zymomonas mobilis. For every test case, MC^3 checks for the dead-end metabolites, zero flux reactions, unsatisfied reversibility and inconsistent irreversibility. Our tool showed several issues with current released models including inconsistent documentation of dead-end metabolites, models with a considerably large number of undocumented zero-reaction fluxes and a number of inconsistently reversible reaction pairs. The MATLAB tool has been placed in the public domain on google code (http://code.google.com/p/mc3/) in April 2012. We hope that this presentation will raise awareness about the need to standardize the documentation of genome-scale stoichiometric models.

(430f) Linear Programming-Based Algorithm for Computing Metabolic Pathways From Genome-Scale Networks

Hyun-Seob Song[1], Noam Goldberg[2], Ashutosh Mahajan[2], Sven Leyffer[2] and Doraiswami Ramkrishna[1], (1)School of Chemical Engineering, Purdue University, West Lafayette, IN, (2)Mathematics and Computer Science Division, Argonne National Laboratory, Argonne, IL

Metabolic pathway analysis provides a useful means for exploring structural and functional properties of complex biological networks. While conventional numerical algorithms scale for a number of EMs running into the millions, genome-scale networks could involve significantly larger numbers, making the identification of EMs substantially more challenging. Alternatively, optimization-based algorithms have been suggested for selective extraction of metabolic pathways using mixed integer linear programming (MILP) (Lee et al. 2000) or integer programming (IP) (de Figueiredo et al. 2009; Rezola et al. 2011). While these methods raise prospects of identifying the full set of pathways for genome-scale networks, the process is still expensive because it involves additional integer variables to exclude previously found solutions. In this work, we present a novel optimization-based algorithm for efficient computation of metabolic pathways. The proposed method computes pathways usinglinear programming (LP) under integer-based constraints that are incorporated in a routine separated from the main optimization problem. Such a strategic partitioning of these two tasks (i.e., optimization solution and constraint design) into separate modules significantly reduces the memory requirement and enhances the computation speed by orders of magnitude. The developed algorithm may be useful in a wide range of applications including (i) metabolic pathway analysis, (ii) flux variability analysis, and (iii) network-based dynamic metabolic modeling (such as lumped hybrid cybernetic model (Song and Ramkrishna 2010; Song and Ramkrishna 2011).

References:

de Figueiredo LF, Podhorski A, Rubio A, Kaleta C, Beasley JE, Schuster S, Planes FJ. 2009. Computing the shortest elementary flux modes in genome-scale metabolic networks. Bioinformatics 25(23):3158-3165.

Lee S, Phalakornkule C, Domach MM, Grossmann IE. 2000. Recursive MILP model for finding all the alternate optima in LP models for metabolic networks. Computers & Chemical Engineering 24(2-7):711-716.

Rezola A, de Figueiredo LF, Brock M, Pey J, Podhorski A, Wittmann C, Schuster S, Bockmayr A, Planes FJ. 2011. Exploring metabolic pathways in genome-scale networks via generating flux modes. Bioinformatics 27(4):534-540.

Song HS, Ramkrishna D. 2010. Prediction of Metabolic Function From Limited Data: Lumped Hybrid Cybernetic Modeling (L-HCM). Biotechnology and Bioengineering 106(2):271-284.

Song HS, Ramkrishna D. 2011. Cybernetic Models Based on Lumped Elementary Modes Accurately Predict Strain-Specific Metabolic Function. Biotechnology and Bioengineering 108(1):127-140.

(430g) Thermodynamically Constrained Flux Analysis of Actinobacillus Succinogenes

Liliana Angeles-Martinez[1], Michael Binns[1] and Constantinos Theodoropoulos[2], (1)Chemical Engineering and Analytical Science, University of Manchester, Manchester, United Kingdom, (2)School of Chemical Engineering and Analytical Science, University of Manchester, Manchester, United Kingdom

The efficient production of platform chemicals from biofuel byproducts is important for the sustainability of biorefineries as well as for the successful deviation from a fossil fuel-based economy. Biocatalysis is an attractive solution for such conversions to platform chemicals. We have extensively studied the bioconversion of crude glycerol from biodiesel production into succinic acid with the use of *Actinobacillus succinogenes* [1,2].

Metabolic analysis of organisms such as *A. succinogenes* [3] and *Escherichia coli*[4] provides target information for manipulating these systems at the genomic and environmental level in order to enhance yields, giving higher production rates, lower byproduct formation and lower costs when they are used for industrial fermentation.

Different stoichiometry-based methodologies have been proposed for the metabolic analysis of a microorganism [5,6], commonly Flux Balance Analysis (FBA) is used for the prediction of theoretical maximum fluxes through the reactions under certain environmental conditions. However, the predicted flux distribution will normally contain a number of thermodynamically infeasible fluxes. A new methodology called Thermodynamic Flux Analysis of Metabolic Systems (TFAMS) for the optimization of fluxes and metabolites' concentrations subject to mass balance and thermodynamic constraints is implemented, which allows the estimation of thermodynamically feasible ranges of fluxes narrowing the solution space greatly compared with classical FBA, with the remaining uncertainty being due to the unknown mechanism of regulation, diffusion effects, experimental error, etc.

As it is known, a reaction is thermodynamically feasible in a certain direction when the Gibbs free energy change ($\Delta_r G$) is negative. The value of $\Delta_r G$ is affected by environmental conditions such as temperature (T), ionic strength (I), pH, and the reactant's concentrations [7,8]. Here we have used experimental fluxes [1] and applied TFAMS to evaluate the impact of T, I, pH, and even the volume exclusion effects on the thermodynamic constraints and hence their effect on the prediction of flux distribution for succinic acid production from glycerol using *A. succinogenes*.

A flux through a reaction may be zero if it is either 1) blocked, when the enzyme that catalyzes it is absent and therefore the value of the Gibbs free energy of the reaction ($\Delta_r G$) is not taken in account, or 2) at equilibrium, when $\Delta_r G$is equal to zero. Unlike the existing Thermodynamic-based Metabolic Flux Analysis (TMFA) [9], TFAMS considers both cases: blocked and thermodynamic equilibrium for reactions with zero flux. Hence our new method can compute tighter and therefore more accurate flux ranges than previous methods, also it can be easily applied to large systems.

Our more accurate flux calculations are used to clarify the results of metabolic control analysis in previous studies [3] which relied on the classical FBA. This gives a better picture of how to enhance the organisms performance through specific genetic enhancements.

References

1. Vlysidis, A., Binns, M., Webb, C., and Theodoropoulos, C. 2011. Glycerol utilisation for the production of chemicals: conversion to succinic acid, a combined experimental and computational study. Biochemical Engineering Journal, 58-59: 1-11.
2. Vlysidis A., Binns M., Webb C., Theodoropoulos C. 2011. A techno-economic analysis of biodiesel biorefineries: Assessment of integrated designs for the co-production of fuels and chemicals. Energy, 36: 4671-4683.

3. Binns, M., Vlysidis, A., Webb, C., Theodoropoulos, C., de Atauri, P., and Cascante, M. 2011. Glycerol metabolic conversion to succinic acid using Actinobacillus succinogenes: a metabolic network-based analysis. Computer Aided Chemical Engineering 29:1421-1425
4. Angeles-Martinez, L., Binns, M., Theodoropoulos, C., de Atauri, P., and Cascante, M. 2012. Thermodynamically constrained Flux and Control analysis of *Escherichia. Coli.* Manuscript in preparation.
5. Llaneras, F., and Pico, J. 2007. A procedure for the estimation over time of metabolic fluxes in scenarios where measurements are uncertain and/or insufficient. BMC Bioinformatics 8:421
6. Klamt, S., and Stelling, J. 2003. Two approaches for metabolic pathway analysis? Trends in Biotechnology 21:64-69.
7. Alberty, R.A. 2003. Thermodynamics of Biochemical Reactions. Wiley-Interscience. The USA. pp 1-17, 35-88.
8. Stephanopoulos, G.N., Aristidou, A.A., and Nielsen, J. 1998. Metabolic Engineering. Principles and Methodologies. Academic Press. The USA. pp 629-694.
9. Henry, C.S., Broadbelt, L.J., and Hatzimanikatis, V. 2007. Thermodynamics-Based Metabolic Flux Analysis. Biophysical Journal 92:1792-1805.

(430h) From Genomes to Tissue-Level Metabolic Models- A Step towards Pathogenesis and Personalized Medicine

Anu Raghunathan[1], Mridula Prasad[2] and Riya Uthup[2], (1)Chemical Engineering, National Chemical Laboratory, Pune, India, (2)National Chemical Laboratory, Pune, India

Renewed interest in metabolic changes in cancer has been spurred on due to a decreasing number of newly released anticancer drugs. In constraint-based analysis, functional states of network can be limited by various physicochemical constraints. Integration of OMICS data into a genome scale network of human metabolism can help to predict the tissue specific metabolic activity in normal and tumor cells. Gene expression data delineates components in a gene list for a specific tissue and suggests information about gene activity for a particular condition. In our modeling approach, we use gene expression data to constrain the maximum flux in the previously published model (Recon1) for human metabolism. Using this approach we are able to predict changes in flux capacity of every enzyme (coded by a gene) in the human metabolic network. We experimentally study changes in metabolism in A549 lung adenocarcinoma cells using phenotypic metabolite arrays and further use mass spectrometry to identify a spectrum of metabolites as predicted by the *in silico* cell at high growth rates.

We have identified diagnostic/prognostic measures (like Positron Emission Tomography with glucose, amino acid analogs) when combined with patient blood plasma microarray data and allowed us to predict clinically relevant tumor parameters like growth and response to therapy including stereotactic radiation therapy. A better understanding of links between cell metabolism and growth in the tumor cell can thus be achieved using this systems approach. Applied to patient models this can facilitate identification of biomarker panels and unique signatures and help drive individualized therapy and personalized medicine.

(497a) The Membrane Environment Can Promote or Suppress Bistability in Cell Signaling Networks

Steven M. Abel, Chemical Engineering, Massachusetts Institute of Technology, Cambridge, MA and Arup K. Chakraborty, Chemical Engineering, Chemistry, and Biological Engineering, Massachusetts Institute of Technology, Cambridge, MA; Ragon Institute of MGH, MIT, & Harvard

Many key biochemical reactions that mediate signal transduction in cells occur at the cell membrane, yet how the two-dimensional membrane environment influences the collective behavior of signaling networks is poorly understood. In an effort to gain insight into the role of the membrane in signal transduction, we use computational methods that explicitly account for stochastic fluctuations to study two topologically different signaling pathways that exhibit bistability: the distributive enzymatic modification of a protein at multiple sites and the positive feedback-mediated activation of a protein. The choice of these networks is motivated by membrane-proximal signaling events in T cells, which act as cellular detectors of pathogens and orchestrate adaptive immunity. In both networks we find that confining proteins to a membrane-like environment can markedly alter the emergent dynamics. The signaling motifs are influenced by membrane features including reduced protein mobility, increased protein concentration, and altered spatiotemporal correlations between pairs of enzyme and substrate molecules. For the distributive protein modification network, increased protein concentration promotes bistability, while lower mobility and membrane-enhanced spatiotemporal correlations suppress bistability. For the positive feedback-mediated activation network, confinement to a membrane environment enhances protein activation, and spatially localized concentration fluctuations can result in the formation and growth of regions with high concentrations of active proteins. By comparing the behavior of the two networks, it is seen that the influence of the membrane on signaling can be qualitatively different for signaling motifs with different network topologies.

(497b) Prediction of Therapeutic Microrna Based On Human Metabolic Network

Ming Wu, Computer Science and Engineering, Michigan State University, East Lansing, MI and **Christina Chan**, Chemical Engineering and Materials Science, Michigan State University, East Lansing, MI

A global decrease in microRNA (miRNA) levels and alterations of miRNA regulation has been observed in human cancers, which suggests miRNAs could have an intrinsic function in tumor suppression. Thus, there is interest in the cancer research community to use miRNA as alternative therapeutic targets. Novel approaches have been developed to target and deliver miRNAs to mammalian system that can successfully inhibit tumor cell proliferation and induced cancer-specific apoptosis. Although miRNAs could be good alternative targets for cancer treatment, it has been difficult to identify which miRNA to target for a particular type of cancer due to our limited knowledge of the regulatory roles of miRNAs in cancer. In this study we propose a novel approach to tackle this challenge, by integrating miRNA target prediction, metabolic modeling and context-specific gene expression data to predict therapeutic miRNAs that can reduce the growth of cancer. Our analysis focuses on the human metabolic system, since abnormal metabolic functions are known to be involved in supporting tumor growth and proliferation. We hypothesized that miRNAs could be implicated in the metabolic regulation of cancer, and developed a novel approach to simulate context dependent metabolic states upon perturbation of gene expression (e.g. induced by miRNA). We incorporate tissue/cancer-specific gene expression information with the generic human metabolic network to reconstruct a context-specific cancer metabolic network, in which miRNAs are perturbed *in silico* to predict their function on metabolic processes and cancer cell growth. Compared with experimental evidences collected from the literature, our approach achieves 89 percent accuracy in predicting miRNAs that can suppress metastasis and progression of human hepatocellular carcinoma (HCC). Our methodology predicts how perturbations in miRNAs impact the cancer metabolic states, thereby suggesting novel mechanisms by which miRNAs regulate cancer cell growth through modulating metabolic functions. We also applied our system to human breast cancer to identify therapeutic miRNA targets for breast cancer.To the best of our knowledge, this is the first computational approach implemented to predict therapeutic miRNAs for human cancer based upon their functional role in cancer.

Minimal reaction network for bistability in the MAPK signalling cascade

O. Hadač, I. Schreiber, M. Přibyl and [1,2]S. Y. Shvartsman

Department of Chemical Engineering, Institute of Chemical Technology Prague, Prague, Czech Republic, E-mail: Otto.Hadac@vscht.cz
[1] Department of Chemical Engineering, Princeton University, New Jersey, United States of America
[2] Lewis-Sigler Institute for Integrative Genomics, Princeton, New Jersey, United States of America

Abstract

Mitogen-activated protein kinases (MAPKs) are important signal transducing enzymes, unique to eukaryotes, that are involved in many pathways of cellular regulation. The MAPK cascade is one of the most studied signalling biochemical pathways. This biochemical network involves many proteins, including MAPKs, which communicate by adding phosphate groups to a neighbouring protein, which in turn acts as an "on" or "off" switch. The cascade is well conserved and always consists of a MAPK kinase kinase (MAPKKK), a MAPK kinase (MAPKK), and a MAPK. The work examined the MAPK cascade described by the Ferrell-Huang model. We used methods of the stoichiometric networks analysis and nonlinear dynamics to reduce the network and obtain minimal reaction network for bistability in the MAPK signalling cascade. Finally, we studied properties of the core model for bistability.

Keywords: Bistability; MAPK cascade; Stoichiometric Network Analysis.

1 Introduction

Mitogen-activated protein kinases (MAPKs) comprise a family of protein-serine/threonine kinases, which participate in signal transduction pathways that control intracellular events in eukaryotes (Pearson et al., 2001). MAPKs are evolutionarily conserved enzymes connecting cell surface receptors to critical regulatory targets such as proliferation, differentiation, and apoptosis within cells (Chang and Karin, 2001). Although biological responses associated with MAPK signalling are highly varied, the basic structure of the MAPK cascade is well conserved. MAPK cascades usually consist of three levels, where the activated kinase at each level phosphorylates the kinase at the next level down the cascade (Markevich et al., 2004). The MAPK cascade always consists of a MAPK kinase kinase (MAPKKK), a MAPK kinase (MAPKK), and a MAPK. MAPKKKs activate MAPKKs by phosphorylation at two conserved serine residues and MAPKKs activate MAPKs by phosphorylation at conserved threonine and tyrosine residues. These signalling pathways are characterized by high levels of structural and parametric uncertainty (Qiao et al., 2007).

The first model of the MAPK cascade was developed by Huang and Ferrell Jr (1996). A statistical exploration of this model in high dimensional parameter space discovered a large region of oscillations and explained their emergence from single-stage bistability (Qiao et al.,

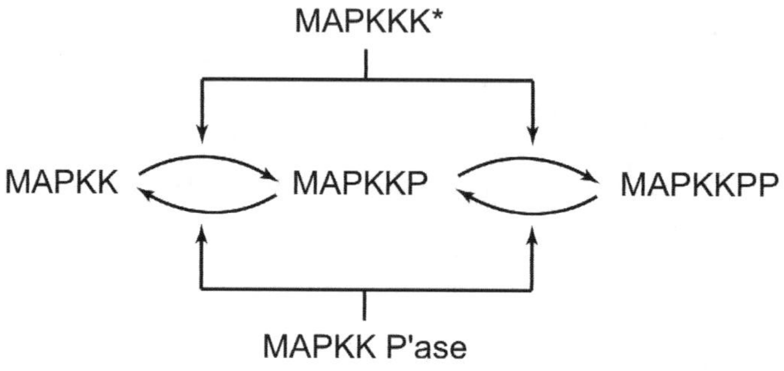

Figure 1: Analyzed MAPK cascade structure from the Ferrell-Huang model.

2007). In this paper we use the stoichiometric networks analysis (SNA) (Clarke, 1980) to decompose a system involving first of the two double-phosphorylation cascades into irreducible or elementary subnetworks (ESs). Then we identify those ESs that are potentially sources of nontrivial dynamical instabilities leading to bistability. We further simplify the topology of the MAPK reaction network into the smallest network which still preserves bistability. This network can be further examined analytically using both convex parametrization employed by the SNA and classical kinetic parameters (i.e., rate coefficients) to determine multiple steady states and their stability.

2 Model of MAPK cascade

This work examines a subsystem of MAPK cascade described by the Ferrell-Huang model (Qiao et al., 2007). It consists of the MAPKK double phosporylation/double dephosphorylation cycle, see FIGURE 1. The activated kinase kinase MAPKKK* phosphorylates two substrates: the MAPKK and its single phosphorylated form MAPKKP. The phosphatase MAPKKP'ase dephosphorylates both double and single phosphorylated forms of the MAPK kinase (MAPKKPP, MAPKKP). Assuming that each enzymatic reaction obeys Michaelis-Menten kinetics, the subsystem involves 12 chemical reactions summarized in TABLE 1. For clarity and compact form of equations, we introduced a new notation for chemical species, see TABLE 2.

Time evolution of the concentrations of the species involved is given by the following mass balance equations:

$$\frac{\mathrm{d}[\mathrm{A}^*]}{\mathrm{d}t} = -v_1 + v_2 + v_3 - v_4 + v_5 + v_6, \tag{1a}$$

$$\frac{\mathrm{d}[\mathrm{B}]}{\mathrm{d}t} = -v_1 + v_2 + v_{12}, \tag{1b}$$

$$\frac{\mathrm{d}[\mathrm{B}_1]}{\mathrm{d}t} = v_3 - v_4 + v_5 + v_9 - v_{10} + v_{11}, \tag{1c}$$

$$\frac{\mathrm{d}[\mathrm{B}_2]}{\mathrm{d}t} = v_6 - v_7 + v_8, \tag{1d}$$

$$\frac{\mathrm{d}[\mathrm{E}_3]}{\mathrm{d}t} = -v_7 + v_8 + v_9 - v_{10} + v_{11} + v_{12}, \tag{1e}$$

$$\frac{\mathrm{d}[\mathrm{A}^*\mathrm{B}]}{\mathrm{d}t} = v_1 - v_2 - v_3, \tag{1f}$$

$$\frac{\mathrm{d}[\mathrm{A}^*\mathrm{B}_1]}{\mathrm{d}t} = v_4 - v_5 - v_6, \tag{1g}$$

$$\frac{\mathrm{d}[\mathrm{E}_3\mathrm{B}_1]}{\mathrm{d}t} = v_{10} - v_{11} - v_{12}, \tag{1h}$$

$$\frac{\mathrm{d}[\mathrm{E}_3\mathrm{B}_2]}{\mathrm{d}t} = -v_7 + v_8 + v_9. \tag{1i}$$

Concentrations of other species (ATP, water) are nearly constant and can be included in the rate coefficients k_j. The reaction terms v_j strictly follow mass action kinetics implied by TABLE 1. Since the total amounts of all three enzymes are conserved in the studied system, mass balance equations (1) must obey the following conservation constraints,

$$\mathrm{A}_{tot} = [\mathrm{A}^*] + [\mathrm{A}^*\mathrm{B}] + [\mathrm{A}^*\mathrm{B}_1], \tag{2a}$$

$$\mathrm{B}_{tot} = [\mathrm{B}] + [\mathrm{B}_1] + [\mathrm{B}_2] + [\mathrm{A}^*\mathrm{B}] + [\mathrm{A}^*\mathrm{B}_1] + [\mathrm{E}_3\mathrm{B}_1] + [\mathrm{E}_3\mathrm{B}_2], \tag{2b}$$

$$[\mathrm{E}_3]_{tot} = [\mathrm{E}_3] + [\mathrm{E}_3\mathrm{B}_1] + [\mathrm{E}_3\mathrm{B}_2]. \tag{2c}$$

3 Results

The reaction network of the MAPK cascade involves nine chemical species ($n = 9$), namely A^*, B, B_1, B_2, E_3, $\mathrm{A}^*\mathrm{B}$, $\mathrm{A}^*\mathrm{B}_1$, $\mathrm{E}_3\mathrm{B}_1$, and $\mathrm{E}_3\mathrm{B}_2$, which occur in 12 elementary reaction steps with kinetics specified in TABLE 1. The evolution equations of the network corresponding to Eqs. (1a)–(1i) which is given by from product of the stoichiometric matrix and the reaction rate vector,

$$\frac{\mathrm{d}\mathbf{x}}{\mathrm{d}t} = \dot{\mathbf{x}} = \mathbf{S}\,\mathbf{v}(\mathbf{x}, \mathbf{k}). \tag{3}$$

All steady-state reaction rates $\mathbf{v}^0$ are determined by

$$\dot{\mathbf{x}} = \mathbf{S}\,\mathbf{v}^0 = \mathbf{0}, \tag{4}$$

Table 1: The reaction mechanism of MAPK cascade

No.	Reaction	Reaction rate
(1)	$B + A^* -> A^*B$	$v_1 = k_1[A^*][B]$
(2)	$A^*B -> B + A^*$	$v_2 = k_2[A^*B]$
(3)	$A^*B -> A^* + B_1$	$v_3 = k_3[A^*B]$
(4)	$B_1 + A^* -> A^*B_1$	$v_4 = k_4[A^*][B_1]$
(5)	$A^*B_1 -> B_1 + A^*$	$v_5 = k_5[A^*B_1]$
(6)	$A^*B_1 -> B_2 + A^*$	$v_6 = k_6[A^*B_1]$
(7)	$B_2 + E_3 -> E_3B_2$	$v_7 = k_7[E_3][B_2]$
(8)	$E_3B_2 -> B_2 + E_3$	$v_8 = k_8[E_3B_2]$
(9)	$E_3B_2 -> B_1 + E_3$	$v_9 = k_9[E_3B_2]$
(10)	$B_1 + E_3 -> E_3B_1$	$v_{10} = k_{10}[E_3][B_1]$
(11)	$E_3B_1 -> B_1 + E_3$	$v_{11} = k_{11}[E_3B_1]$
(12)	$E_3B_1 -> B + E_3$	$v_{12} = k_{12}[E_3B_1]$

Table 2: List of symbols standing for chemical species

Symbol	Chemical species
A^*	MAPKKK*
B	MAPKK
B_1	MAPKKP
B_2	MAPKKPP
E_3	MAPKKP'ase
A^*B	MAPKKK*MAPKK
A^*B_1	MAPKKK*MAPKKP
E_3B_1	MAPKKP'aseMAPKKP
E_3B_2	MAPKKP'aseMAPKKPP

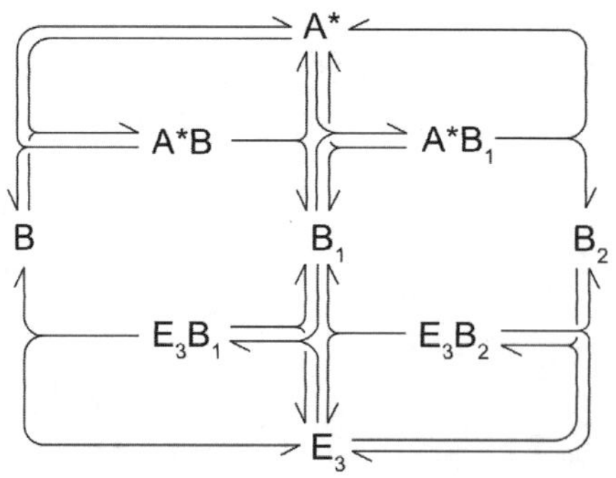

Figure 2: Network diagram of the second MAPK cascade.

and can be expressed as a convex combination of elementary subnetworks $\mathbf{E}_i$,

$$\mathbf{v}^0 = \mathbf{E}\,\mathbf{j}\,, \tag{5}$$

where $\mathbf{j}$ involves coefficients of any convex combination. By examining principal minors of the matrix $-\mathbf{S}\,\mathrm{diag}(\mathbf{Ej})\,\underline{\boldsymbol{\kappa}}^{\mathrm{T}}$, where $\{\kappa_{ij}\} = \left\{\dfrac{\partial \ln v_j}{\partial \ln x_i}\right\}$, instability of any subnetwork can be predicted (Clarke, 1980) prior specifying rate coefficients and/or steady state concentrations.

We apply this stability analysis to all elementary subnetworks and a set of all 2-combinations of elementary subnetworks. Results indicate that all elementary subnetworks $\mathbf{E}_i$ are stable. However, a combination of two subnetworks $\mathbf{E}_i$ is found to possess predicts an instability. In the next step, steady state concentrations are chosen properly and inspection of the eigenvalues of Jacobi matrix clarifies that the instability occurs via saddle-node bifurcation which leads to bistability. In order to identify the topological core of bistability, we made several simplifying steps resulting in the simplest bistable network, which involves five reactions, see FIGURE 3 and TABLE 3. Any attempt to further reduce this network causes that the network only possesses a single stable steady state.

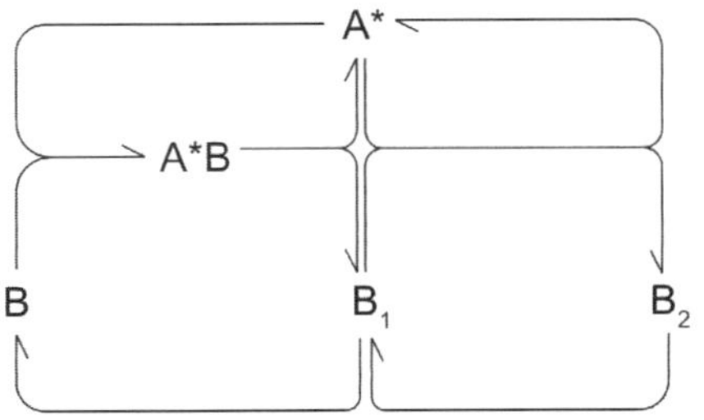

Figure 3: Network diagram of the simplest mechanism of MAPK cascade exhibiting bistability.

3.1 Core bistable network

The *simplest bistable network* in FIGURE 3 forms the core that is responsible for the MAPK bistability. The network involves five chemical species ($n = 5$), which occur in five reaction steps (1)–(5), see TABLE 3. The mass balances of all components of the core subnetwork read:

Table 3: The simplest bistable mechanism of MAPK cascade

No.	Reaction	Reaction rate
(1)	$B + A^* -> A^*B$	$v_1 = k_1[A^*][B]$
(2)	$A^*B -> A^* + B_1$	$v_2 = k_2[A^*B]$
(3)	$B_1 + A^* -> B_2 + A^*$	$v_3 = k_3[A^*][B_1]$
(4)	$B_2 -> B_1$	$v_4 = k_4[B_2]$
(5)	$B_1 -> B$	$v_5 = k_5[B_1]$

$$\frac{dx_1}{dt} = k_2x_5 - k_1x_1x_2, \quad (6a)$$
$$\frac{dx_2}{dt} = -k_1x_1x_2 + k_5x_3, \quad (6b)$$
$$\frac{dx_3}{dt} = k_2x_5 - k_3x_1x_3 + k_4x_4 - k_5x_3, \quad (6c)$$
$$\frac{dx_4}{dt} = k_3x_1x_3 - k_4x_4, \quad (6d)$$
$$\frac{dx_5}{dt} = k_1x_1x_2 - k_2x_5. \quad (6e)$$

where x_1, x_2, x_3, x_4, and x_5 stand for [A^*], [B], [B_1], [B_2], and [A^*B], respectively. Since rank(**S**) = 3 there must exist two conservation constraints for mass balance of the kinase A^* and the substrate B,

$$A_{tot} = x_1 + x_5 \,, \quad (7a)$$
$$B_{tot} = x_2 + x_3 + x_4 + x_5 \,. \quad (7b)$$

The steady states of Eqs. (6a)-(6e) can be expressed analytically by substituting for x_2 and x_5 from the conservation equations (7a)–(7b). We obtain a cubic equation

$$k_1k_2k_3x_1^3 + (k_1k_4k_5 + k_1k_2k_4 - k_1k_2k_3A_{tot})x_1^2 + (k_2k_4k_5 - k_1k_4A_{tot}(k_2 + k_5) + k_1k_4k_5B_{tot})x_1 - k_2k_4k_5A_{tot} = 0 \,, \quad (8)$$

that yields either three real roots, corresponding to the region of bistability, or one real and two complex conjugate roots, corresponding to the region of a single steady state. The steady state values for the other variables can be simply determined from

$$x_3 = k_2(A_{tot} - x_1)/k_5 \,, \quad (9a)$$
$$x_4 = k_2k_3x_1(A_{tot} - x_1)/(k_4k_5) \,. \quad (9b)$$

Linear stability of the steady states can be then examining by means the eigenvalues of the Jacobi matrix of Eqs. (6a)–(6e).

4 Conclusions

We used the stoichiometric network analysis to decompose the subsystem of the MAPK network shown in FIGURE 2 into elementary subnetworks (ESs). We identified those ESs that are potentially sources of dynamical instabilities leading to bistability and further simplified the topology of the MAPK network into the smallest network which still preserves bistability. This network was examined analytically to determine multiple steady states and their stability.

The simplest bistable network in FIGURE 3 can be decomposed into two subnetworks, which are reflecting the first and second phosphorylation/dephosphorylation steps. The first phosphorylation/dephosphorylation step involves reaction steps (1), (2) and (5) from TABLE 3, the second phosphorylation/dephosphorylation step involves reaction steps (3) and (4). The stability analysis indicates that each subnetwork is stable, but under suitable conditions their combination may produce an unstable steady state of a saddle type.

We also found out that at stable steady states both subnetworks contribute very unequally to the whole network whereas at the saddle their contribution is much more even. In addition, the phosphorylated steady state, characterized by high concentration of B_2 (x_4), is marked by the dominance of the second subnetwork related to the second phosphorylation/dephosphorylation steps. In contrast, the dephosphorylated steady state, characterized by a high concentration of B (x_2), is dominated by the first phosphorylation/dephosphorylation subnetwork.

5 Acknowledgements

This work was supported by the grant MŠM ČR KONTAKT ME 10036.

References

G. Pearson, F. Robinson, T. B. Gibson, B. E. Xu, M. Karandikar, K. Berman, and M. H. Cobb, Endocrine Reviews **22**, 153 (2001).

L. Chang and M. Karin, Nature **410**, 37 (2001).

N. I. Markevich, J. B. Hoek, and B. N. Kholodenko, Journal of Cell Biology **164**, 353 (2004).

L. Qiao, R. B. Nachbar, I. G. Kevrekidis, and S. Y. Shvartsman, PLoS Computational Biology **3**, 1819 (2007).

C. Y. F. Huang and J. E. Ferrell Jr, Proceedings of the National Academy of Sciences of the United States of America **93**, 10078 (1996).

B. L. Clarke, Advances in Chemical Physics **43**, 1 (1980).

(497d) Modeling the Dynamics of Acute Phase Protein Expression in HepG2 Cells Stimulated by IL-6

Jens O.M. Karlsson, Department of Mechanical Engineering, Villanova University, Villanova, PA, Zhaobin Xu, Department of Chemical Engineering , Villanova University, Villanova, PA, Arjun Plakkat, Departments of Chemical Engineering , Pennsylvania State University, University Park, PA, Calvin Hong Li, Department of Mechanical Engineering , Villanova University, Villanova, PA and **Zuyi (Jacky) Huang**, Department of Chemical Engineering, Villanova University, Villanova, PA

Interleukin-6 (IL-6) triggers the human body's acute phase response to trauma or inflammation mainly by stimulation of two signaling pathways: the first pathway involves the activation of the Janus-associated kinases (JAK) and signal transducers and activators of transcription (STAT3), while the second pathway comprises the activation of mitogen-activated protein kinases (MAPK) and CCAAT (cytidine-cytidine-adenosine-adenosine-thymidine)/enhancer binding protein β (C/EBPβ) (Heinrich et al., 2003). The expression of genes of certain acute phase proteins, such as albumin, haptoglobin, and fibrinogen, is regulated by the transcription factors associated with IL-6 signaling, i.e., nuclear STAT3 dimer and C/EBPβ (Alam et al., 1992; Fuller and Zhang, 2001; Ruminy et al., 2001). The expressed acute phase proteins in turn participate in the human body's response to trauma or inflammation. Hence, an accurate mathematical model for predicting the dynamics of acute phase protein expression in hepatocytes can facilitate the treatment of trauma or inflammation. Although mathematical models for IL-6 signaling pathways have previously been developed (Huang et al., 2009; Moya et al., 2011; Singh et al., 2006), reactions that describe the expression of acute phase proteins have not been included in these models. To address this deficiency, a recent model of IL-6 signaling (Moya et al., 2011) is extended in this work to predict the dynamics of acute phase protein expression in IL-6-stimulated HepG2 cells (a human hepatoma cell line). A sensitivity analysis is then performed to identify reactions in IL-6 signaling that play an important role in regulation of the dynamics of acute phase protein expression.

The starting point for the model used in this work is the IL-6 signal transduction model presented in Moya et al., 2011. This model is based upon the model structure proposed by Singh et al., 2006, whereas the dynamic model of Jak-STAT signaling is adopted from Yamada et al., 2003, and parts of the detailed kinetic model of MAPK signaling proposed by Schoeberl et al., 2002 are also used. In the present work, reactions that describe the expression of haptoglobin, fibrinogen, and albumin were added into the IL-6 signaling model. C/EBPβ, nuclear STAT3 dimer, and C/EBPα are the transcription factors regulating the transcription of haptoglobin (Alam et al., 1992), fibrinogen (Zhang and Fuller, 2001), and albumin (Alam et al., 1992; Ruminy et al., 2001), respectively. For example, C/EBPβ down-regulates the expression of albumin by inhibiting the activation of C/EBPα (Ruminy et al., 2001). Following this approach, a system of 75 coupled ordinary differential equations was developed based on mass action kinetics and Michaelis-Menten kinetics. Experimental data for HepG2 cells stimulated by 2 nM IL-6 continuously for seven hours (Karlsson et al., 1998) were used to estimate the values of parameters involved in the newly added reaction equations. The extended model was then validated against independent experimental data for HepG2 cells under three-hour exposure to three different IL6 concentrations (Karlsson et al., 1998). Experimental data for a pulse-chase stimulation with 2 nM IL-6 (Karlsson et al., 1998) were used to further validate the extended model. In addition, the predicted steady-state accumulation rates of fibrinogen and albumin for various IL-6 concentrations were compared to the experimental data reported by Heinrich et al., 1990, and found to be consistent with the published measurements. Finally, a sensitivity analysis was performed to identify reactions in IL-6 signaling pathway that play an important role in regulating the expression of haptoglobin, fibrinogen, and albumin, respectively. Binding of STAT3C to $(IL6–gp80–gp130–JAK)_2^*$ and expression of SOCS3 were identified as essential steps for regulating the secretion rate of fibrinogen, while the conversion of Ras–GDP to Ras–GTP, the unbinding process of $SHP2^*$–Grb2–SOS with $(IL6–gp80–gp130–JAK)_2^*$, and the deactivation of Raf^* by phosphatase Phospl were determined as the key processes that regulate the dynamics of haptoglobin and albumin secretion from HepG2 cells.

References

Alam T., An M.R., Papaconstantinou J., (1992). Differential expression of three C/EBP isoforms in multiple tissues during the acute phase response. The Journal of Biological Chemistry, 267(8), 5021-5024.

Heinrich, P. C., Behrmann, I., Haan, S., Hermanns, H. M., Muller-Newen, G., Schaper, F., (2003). Principles of interleukin (IL)-6-type cytokine signaling and its regulation. Biochemical Journal, 374, 1-20.

Heinrich, P.C., Castell, J.V., Andus, T., (1990). Interleukin-6 and the acute phase response. Biochemical Journal, 265(3), 621-636.

Huang, Z., Moya, C., Cheng, P., Jayaraman, A., Hahn, J., (2009). In silico investigation of IL-6 and IL-10 signaling in steatosis. Proceedings Foundations of Systems Biology in Engineering (FOSBE), Denver, Colorado, 28-31.

Karlsson, J.O.M., Yarmush, M.L., Toner, M., (1998). Interaction between heat shock and interleukin 6 stimulation in the acute-phase response of human hepatoma (HepG2) cells. Hepatology, 28(4), 994-1004.

Moya, C., Huang, Z., Cheng, P., Jayaraman, A., Hahn, J., (2011). Investigation of IL-6 and IL-10 signaling in steatosis via mathematical modeling. IET Systems Biology, 5(1), 15-26.

Ruminy, P., Gangneux, C., Claeyssens, S., Scotte, M., Daveau, M., Salier, J.-P, (2001). Gene transcription in hepatocytes during the acute phase of a systemic inflammation: from transcription factors to target genes. Inflammation Research, 50, 383-390.

Schoeberl, B., Eichler-Jonsson, C., Gilles, E. D., Muller, G., (2002). Computational modeling of the dynamics of the MAP kinase cascade activated by surface and internalized EGF receptors. Nature Biotechnology, 20, 370-375.

Singh, A. K., Jayaraman, A., Hahn, J., (2006). Modeling regulatory mechanisms in IL-6 signal transduction in hepatocytes. Biotechnology & Bioengineering, 95(5), 850-862.

Yamada, S., Shionoa, S., Joob, A., Yoshimura, A., (2003). Control mechanism of JAK/STAT signal transduction pathway. FEBS Letters, 534, 190-196.

Zhang, Z., Fuller, G.M., (2000). Interleukin 1β inhibits interleukin 6–mediated rat γ fibrinogen gene expression. Blood, 96(10), 3466-3472.

(497e) Resolution of Incoherent Immune Stimuli Through Intracellular Crosstalk

Camila Benaim, Computer Science, ISP, Northwestern University, Evanston, IL, **Yishan Chuang**, Chemical and Biologial Engineering, Northwestern University, Evanston, IL and Joshua N. Leonard, Chemical and Biological Engineering, Chemistry of Life Processes Institute, R.H. Lurie Comprehensive Cancer Center, Northwestern University, Evanston, IL

Macrophages play a critical role in maintaining the balance between homeostasis and protective inflammation by adopting either immunostimulatory (M1) or immunosuppressive (M2) phenotypes. Tumors and pathogens manipulate this balance by inducing the production of immunosuppressive stimuli in immune microenvironment, which shifts macrophages into a tumor or pathogen-promoting M2 phenotype. In other experimental work, we discovered that sustained exposure of macrophages to the canonical pro-M2 stimulus IL-10 promotes maintenance of the M2 phenotype even in the presence of IL-12, a stimulus that previously reported to convert tumor-associated M2 cells to a M1 phenotype. This observation contradicts a dominant conceptual model in the field that had motivated clinical administration of IL-12 to cancer patients to overcome local immune dysfunction. In order to investigate potential mechanisms that could mediate this new phenomenon, we developed mechanistic computational models of the IL-10 and IL-12 signaling pathways and key downstream gene circuits involved in cross-regulation. Briefly, IL-10 induces signaling via the transcription factor STAT3, which promotes the production of immunosuppressive mediators as well as suppressors of cytokine signaling 1 and 3 (SOCS1, SOCS3). SOCS1 inhibits signaling downstream of both IL-10 and IL-12, but SOCS3 inhibits signaling downstream from IL-12 but not IL-10. On the other hand, IL-12 signals through STAT4, which promotes the production of immunostimulatory mediators and SOCS1. Therefore, SOCS1 and SOCS3 potentially mediate asymmetric macrophage fate determination through crosstalk between pathways and gene circuits induced by IL-10 and IL-12. A challenge in evaluating such a theory is determining whether multiple observations (qualitative, semi-quantitative, and quantitative) may be integrated into a single parsimonious description of this system that is consistent with all observations. To meet this need, here we describe the use of both stochastic and deterministic simulations to interrogate the properties of this important intracellular circuit and to evaluate potential mechanisms by which this key facet of immune dysfunction in chronic disease may be regulated and therapeutically manipulated.

(497f) A Kinetic Platform to Determine the Fate of Nitric Oxide in Bacteria

Jonathan L. Robinson, Chemical and Biological Engineering, Princeton University, Princeton, NJ and Mark P. Brynildsen, Chemical and Biological Engineering, Princenton University, Princeton, NJ

Nitric oxide (NO•) is generated by cells of the innate immune response (*e.g.*, macrophages) to neutralize pathogens. NO• has an extensive biochemical reaction network that includes reactions with iron-sulfur clusters, DNA, thiols, tyrosine residues, and hemes. The fate of NO• inside a pathogen depends on the kinetic competition among its many targets. This competition is governed by microbial parameters such as intracellular metabolite and enzyme concentrations, and is of critical importance as it determines the fitness of an organism in NO•-stressed environments. Many NO• reaction products are difficult or impossible to directly measure (*e.g.*, NO_2•, N_2O_3, $ONOO^-$), and their terminal pathway products are non-unique (*e.g.*, NO_2^-, NO_3^-, N_2O). Therefore, dynamic kinetic models have been developed to predict concentrations of NO• and its products in aqueous solutions and physiological environments, such as inflamed tissue.[1,2] These efforts have laid a foundation for modeling NO• reaction networks, but have focused primarily on non-enzymatic, small-molecule reactions and not addressed the fundamental question of how NO• distributes within microbes. Here, we have constructed a comprehensive kinetic model that unites and extends existing treatments of NO• chemistry to encompass the broad reactivity of NO• in *E. coli*. The incorporation of spontaneous and enzymatic reactions, as well as damage and repair of biomolecules, allows for a detailed analysis of intracellular NO• distribution and its dependence on various microbial parameters. Interestingly, a parametric analysis of our dynamic model identified a dependency in the utility of NO• dioxygenase (Hmp) on the rate of NO• delivery to the cell. Hmp is the major enzyme responsible for aerobic NO• detoxification in *E. coli*, but when NO• is delivered at a rapid rate, even for short periods of time, the distribution and removal of NO• become unaffected by the presence of Hmp. This observation was experimentally confirmed, as an Δ*hmp* mutant exhibited wild-type sensitivity toward rapidly delivered NO•, whereas a considerable increase in sensitivity for Δ*hmp* compared to wild-type was observed at slower delivery rates. Our model serves as a platform to predict the fate of NO• within bacteria and offers a framework to identify novel NO•-potentiating therapeutics.

1. Lancaster, J. R., Jr. *Chem. Res. Toxicol.* **2006**.

2. Lim, C. H. *et al. Chem. Res. Toxicol.* **2008**.

(497g) Decoding Robustness in Integrated Signaling and Gene Regulatory Networks Using a Combination of Global Sensitivity Analysis and Decision Trees

Hirenkumar Makadia, Daniel Baugh Institute for Functional Genomics and Computational Biology, Department of Pathology Anatomy and Cell Biology, Thomas Jefferson University, Philadelphia, PA, James S. Schwaber, Daniel Baugh Institute for Functional Genomics and Computational Biology, Department of Pathology, Cell Biology and Anatomy, Thomas Jefferson University, Philadelphia, PA and Rajanikanth Vadigepalli, Daniel Baugh Institute for Functional Genomics and Computational Biology, Department of Pathology, Anatomy and Cell Biology, Thomas Jefferson University, Philadelphia, PA

The primary objective of our study is to identify the mechanisms through with the robustness is encoded in a complex network of signaling pathways and gene regulatory networks connected through feed-forward and feedback interactions. The complexity of the regulatory networks containing combinatorial/nonlinear interactions, cross-talk, and feedback circuitry requires investigation beyond topological aspects by explicitly analyzing the network dynamics. We present a novel, systematic, and mathematical approach combining global sensitivity analysis and decision trees, which elucidates such nonlinear relationships and identifies the robust as well as sensitive control points in the network. We focused on the robustness of activation for immediate early transcription factors in response to variations in the dynamic profiles of upstream signaling pathways. The variations in signaling kinase activities represents differential function in individual cells and cell phenotypes. We demonstrate our approach on a biological system of key relevance to neuronal adaptive processes dysregulated in the development of hypertension. Our results indicate several aspects of the network robustness that are not intuitable from static topological considerations alone.

We considered the angiotensin II type 1 receptor (AT1R) signaling and gene regulatory network in the brainstem [1]. AT1R network mediates a wide range of autonomic functions in brainstem, one of which is the control of hypertension by regulation of Tyrosine hydroxylase (TH) production. The gene regulatory network activated by AT1R was previously studied for robustness of transcription factor (TF) activation dynamics to variations in the kinetic parameters[1]. This mechanistic model was implemented as a set of ordinary differential equations, and bridges three signaling kinases viz. FRK, ERK and JNK with activation of AP-1 TF family. In the present study, we seek to identify the robustness of AP-1 and TH activation to variations in the dynamics of kinases FRK, ERK and JNK, to understand the decoding of the combinatorial signaling activity at the level of transcription factor dynamics and the consequences on target gene expression.

The kinase temporal profiles were considered as a set of nonlinear exponential functions with six parameters describing their delay and kinetics of activation, peak retention and deactivation. Potential kinase profiles were randomly sampled in two-fold range to explore network dynamics in biologically meaningful range of signaling activities. We followed a global sensitivity analysis approach using Sobol sensitivity indices for a total of eighteen kinase parameters as they affect AP-1 and TH dynamics[2]. We categorized the results based on the presence/absence of delay in activation of three kinases and compared the sensitivities across these groups. We further analyzed the dependencies between the key kinase profile parameters that were significantly affecting AP-1 or TH dynamics using a decision tree approach. This method organizes the key kinase parameter ranges in a hierarchy with different paths in the tree leading to distinct AP-1 or TH dynamical behavior.

We identified key kinase profile parameter that control the AP-1 family of TFs (c-fos:c-jun hetrodimer and c-jun:c-jun homodimer). Although topology of the mechanistic model suggest an influence of FRK during initial phase, global sensitivity analysis shows a surprising result that delay in activation of JNK and ERK as the most influential parameters in affecting downstream AP-1 activity and target TH gene expression dynamics. The peak retention parameter of the JNK shows a biphasic influence with reduced sensitivity (i.e., robustness) at the transition phase in which the composition of AP-1 is switching form a heterodimer to a homodimer. A higher order interaction between

JNK peak retention parameter, FRK deactivation parameter and its time constant controlled the transition phase as demonstrated through their second order sensitivities. The transition period, when the the c-jun:c-jun dimer activity dominates the c-fos:c-jun dimer activity, was a robust phase during which the level of AP-1 or TH was least sensitive to all of the kinase parameters. We interpret this as a network-level constraint that differentiates the state of the network that is dominated by a hetero- vs homo-dimer of AP-1. During the last phase in which the homo-dimer dominated the regulation, the activity of AP-1 and TH gene expression was significantly sensitive to all the parameters of JNK except the delay in activation and corresponding time constant.

The decision tree analysis organized these parameters into a hierarchy, with subsets of kinase profiles leading to distinct activation profiles of AP-1 and TH. These indicate that individual cells or cell phenotypes representing such subsets of kinase activities could lead to distinct downstream network response and contribute towards a heterogeneous tissue scale behavior. The robustness of AP-1 gene regulatory network to a majority of the kinase activity parameters, and differential sensitivity to a select subset, reveal the control points in this system that when dysregulated are likely to lead to an aberrant network response underlying maladaptive autonomic control such as in essential hypertension.

References

[1] Gregory M Miller, Babatunde A Ogunnaike, James S Schwaber, and Rajanikanth Vadigepalli. Robust dynamic balance of AP-1 transcription factors in a neuronal gene regulatory network. *BMC Systems Biology*, 4(1):171, 2010.

[2] Andrea Saltelli. *Global Sensitivity Analysis: The Primer*. John Wiley, March 2008.

DOUBLY PENALIZED APPROACH FOR RECONSTRUCTION OF BIOLOGICAL NETWORKS

Behrang Asadi[a,b], *Daniel M. Tartakovsky*[a,1], *Mano Ram Maurya*[b], *Shankar Subramaniam*[b,c,2]

[a]*Department of Mechanical and Aerospace Engineering,*
University of California, San Diego, La Jolla, CA, USA
[b]*Department of Bioengineering,*
[c]*Department of Chemistry & Biochemistry,*
University of California, San Diego, La Jolla, CA, USA
E-mail: [basadi, dmt, mmaurya, shankar]@ucsd.edu

Reconstruction of biological networks is a crucial step in extracting information from a large volume of experimental data. Various methods have been developed to reconstruct networks from data, each of which possesses its own strength and disadvantages. Examples of network reconstruction techniques include classic optimization-based approaches (e.g., least-squares methods), dimensionality reduction methods (e.g., statistical significance tests combined with either principal components regressions (PCR), or partial least-squares (PLS) [1]), Bayesian networks and hybrid methods (e.g., Linear Matrix Inequalities (LMI) and Least Absolute Shrinkage and Selection Operator (LASSO) [2]). A detailed review of various network-reconstruction methods can be found in [3]. In this work, we present a new method, called Doubly Penalized Linear Absolute Shrinkage and Selection Operator (DPLASSO) for network reconstruction with the intent to combine the beneficial features of a regression and statistical significance testing-based method and a penalized optimization method.

In this approach, we have integrated two distinct methods viz., statistical significance testing of model coefficients and penalized/constrained optimization. Principal component analysis with statistical significance testing acts as a supervisory-level filter to extract the most informative components of the network from a dataset (Layer 1). In the lower level (Layer 2), LASSO with extra weights on the smaller parameters obtained in the first layer is employed to retain the main predictors and to set the small coefficients to zero. Two case studies are used to compare the relative performance of DPLASSO and LASSO in terms of several metrics, such as sensitivity, specificity, accuracy and fractional-error in the estimates of the coefficients. In the first case study, with a synthetic data set, our simulation results show substantial improvements over LASSO for the reconstruction of the network in terms of accuracy and specificity. The second case study uses simulated datasets from a nonlinear model for cell division cycle of fission yeast. This case study illustrates that DPLASSO performs better than LASSO in terms of sensitivity, specificity and accuracy in reconstructing networks.

Acknowledgement

This research was supported by the National Heart, Lung and Blood Institute (NHLBI) grant 5 R33 HL087375-02 and National Science Foundation (NSF) grants DBI-0641037, DBI-0835541 and STC-0939370 (Collaborative) to SS.

[1] Corresponding author; Department of Mechanical and Aerospace Engineering, University of California, San Diego, 9500 Gilman Drive, La Jolla, CA 92093-0411, Phone: (858) 534-1375.

[2] Corresponding author; Department of Bioengineering, University of California San Diego, 9500 Gilman Drive, La Jolla, CA, 92093-0412, Phone: (858) 822-0986.

References

1. Gupta, S., M.R. Maurya, and S. Subramaniam, *Identification of crosstalk between phosphoprotein signaling pathways in RAW 264.7 macrophage cells.* PLoS Comput Biol, 2010. **6**(1): p. e1000654.
2. Tibshirani, R., *Regression shrinkage and selection via the Lasso.* Journal of the Royal Statistical Society Series B-Methodological, 1996. **58**(1): p. 267-288.
3. Maurya, M. and S. Subramaniam, *Computational Challenges in Systems Biology*, in *Systems Biomedicine: Concepts and Perspectives*, E. Liu and D. Lauffenburger, Editors. 2009, Academic Press: SAN DIEGO. p. 177-223.

(564a) Effect of Quorum-Signaling Molecules On Human Epithelial Cells: Implications for Interkingdom Response and Communication

Amin Zargar, Bioengineering, University of Maryland, College Park, College Park, MD; Center for Biosystems Research, College Park, MD and William Bentley, Fischell Department of Bioengineering, University of Maryland, College Park, College Park, MD; Institute for Bioscience and Biotechnology Research, College Park, MD

It has been established that bacteria use quorum sensing molecules to coordinate gene expression in a cell density-dependent manner. This group of cell communication molecules is displayed throughout the bacterial hierarchy influence characteristics such as swarming motility, biofilm formation, virulence, among others. A distinct class of signaling molecules, autoinducer-2, is capable of interspecies communication, and is so prevalent throughout the bacterial domain that it is thought to be a "universal" signaling molecule. While the mechanisms behind autoinducer-2 have been clearly elucidated, the interkingdom relationship between quorum signaling molecules and human cells are not yet understood. This relationship is of particular importance considering the many species of bacteria that constitute the microflora inside the intestine. In this study, we explore the effects of autoinducer-2 on gene expression of human epithelial cells using the next-generation RNA-seq.

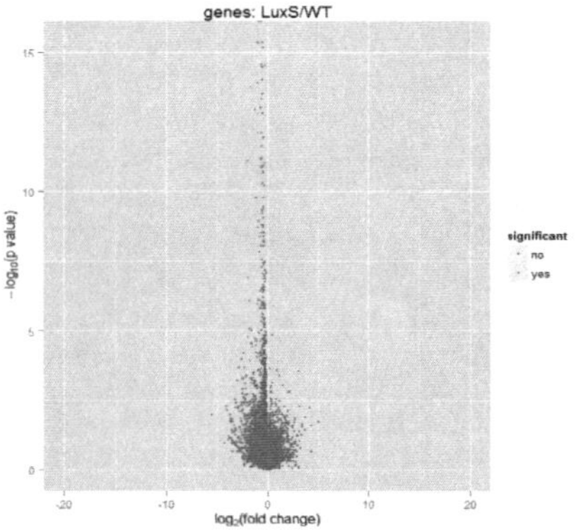

Human epithelial cells, HCT-8, are exposed to BL21, a wild-type bacteria that produce AI-2, and a BL21 LuxS knockout that does not produce AI-2. The RNA is extracted and is sequenced in triplicate on two lanes of the Illumina HiSeq1000 Sequencing System. Over 100 million reads for each condition per biological replicate were analyzed using the Tuxedo Suite (Bowtie, Tophat & Cufflinks). With 90% alignment, a total of 318 differentially expressed genes and 8773 differentially expressed isoforms were found. Gene clustering analysis using DAVID (Database for Annotation, Visualization and Integrated Discovery) led to significant 28 gene clusters and three signaling pathways through Signaling Pathway Impact Analysis.

With the advancements of next generation RNA-Seq technology, sequencing of the eukaryotic transcriptome is now possible with newfound depth and breadth, to determine eukaryotic responses. In this study, transcriptional and post-transcriptional differential processing was explored and biological network paths were inferred to indicate response. Reinforced with this understanding, two-way interkingdom communication can potentially be mediated with genetically engineered bacteria that can act on eukaryotic response mechanisms.

(564b) Freq-Seq: A Method for Rapid, Cost-Effective Allele Frequency Determination in Microbial Populations

Lon Chubiz[1], Ming-Chun Lee[1,2], Nigel F. Delaney[1] and Christopher J. Marx[1,3], (1)Organismic and Evolutionary Biology, Harvard University, Cambridge, MA, (2)Biochemistry, Hong Kong University, Hong Kong, China, (3)FAS Center for Systems Biology, Harvard University, Cambridge, MA

Understanding evolutionary dynamics in microbial systems requires the ability to accurately follow allele frequencies in populations through time. Established methods such as quantitative Sanger sequencing, restriction fragment length polymorphism (RFLP), quantitative PCR (qPCR) and mass spectrometry-based methods (SEQUENOM) provide an array of strategies to interrogate allele frequencies. But these methods are inhibited by both their ease and/or cost of implementation, as well as their respective detection limits. An alternative strategy has been to rely on next generation meta-sequencing of microbial populations to infer relative frequencies of pertinent genotypes. The primary drawbacks to this approach are the lack of uniform sequencing depth required to make quantitative diagnoses of low frequency mutations and inability to perform large scale sample multiplexing.

In contrast, we present a rapid, cost-effective strategy for localized allele frequency detection using the Illumina HiSeq and Genome Analyzer next-generation sequencing platforms called Freq-Seq. The Freq-Seq method utilizes two rounds of PCR using inexpensive user-designed oligonucleotides and a bar-coded adapter system that can be produced in-house at low cost. Likewise, we have developed custom analysis software allowing users to easily extract and interpret experimental results. Both the Freq-Seq allele frequency detection adapter kit and open-source software are now publically available.

We validate and demonstrate the utility of the Freq-Seq system by determining the order of emergence and frequency of beneficial alleles that arose as a microbial population adapted to an engineered metabolic pathway. Our data show that Freq-Seq is capable of detecting allele frequencies accurately below 1% abundance. Likewise, the evolutionary dynamics observed via Freq-Seq allowed us to uncover the existence of large-scale clonal interference during the early stages of adaptation. These results suggest the utility of Freq-Seq as an accessible and cost-effective diagnostic tool in determining the genetic composition of microbial processes and communities.

(564c) Ptm Curator: An Automated Method for the Frequency Analysis of the Experimental and Putative Post-Translational Modification Statistics Contained in the Swiss-Prot Database

George A. Khoury, Richard C. Baliban and Christodoulos A. Floudas, Chemical and Biological Engineering, Princeton University, Princeton, NJ

Post-translational modifications (PTMs) broadly contribute to the recent explosion of proteomic data and possess a complexity surpassing that of protein design [1]. PTMs are the chemical modification of a protein during or after its translation, and have wide effects broadening its range of functionality [2]. Based on previous estimates, it is believed that more than half of proteins are glycoproteins [3]. Whereas mutations can only occur once per position, different forms of post-translational modifications may occur in tandem. With the number and abundances of modifications constantly being discovered, there is no method to readily assess their relative levels.

In this work, we report the relative abundances of each PTM found experimentally and putatively, from high-quality, manually curated, proteome-wide data contained in the Swiss-Prot [4-6] database. We find that at best, less than one-fifth of proteins are glycosylated from the global dataset. We further explore the frequency of the D-amino acids in the database in comparison to the global set. Remarkably we found that D-alanine is the most frequent, which may have implications for the origins of life given alanine's status as the "default" amino acid. Only 837 D-amino acids were found to be contained in the database of over 187,941,074 amino acids [7].

The frequencies can be converted into probabilities and conditional probabilities representing the possibility of identification of a PTM on a peptide or protein. This can help untargeted proteomic assignment to develop probabilistic methods to determine the expected value associated with potential PTM sites on proteins, helping account for uncertainty. Similarly, the statistics can help validate targeted assignment of modifications.

We make available to the academic community a continuously updated ***systems-wide*** resource, the **PTM Curator** [7] (http://selene.princeton.edu/PTMCuration). New features have been added for ease of use. Namely, we populate the UniProt ID numbers associated with each modification and filter the modification statistics by whether they occurred on a prokaryote or eukaryote, as well as provide the modifications occurring on mammalian proteins. Using our Curation method, scientists unambiguously can assess "how many" of each PTM exists.

References:

1. Baliban, R.C., et al., *A Novel Approach for Untargeted Post-translational Modification Identification Using Integer Linear Optimization and Tandem Mass Spectrometry.* Molecular & Cellular Proteomics, 2010. **9**(5): p. 764-779.

2. Walsh, C., *Posttranslational modification of proteins: expanding nature's inventory*2006, Englewood, Colo.: Roberts and Co. Publishers. xxi, 490 p.

3. Apweiler, R., H. Hermjakob, and N. Sharon, *On the frequency of protein glycosylation, as deduced from analysis of the SWISS-PROT database.* Biochimica et Biophysica Acta (BBA) - General Subjects, 1999. **1473**(1): p. 4-8.

4. Bairoch, A. and R. Apweiler, *The SWISS-PROT protein sequence database and its supplement TrEMBL in 2000.* Nucleic Acids Research, 2000. **28**(1): p. 45-48.

5. Jung, E., et al., *Annotation of glycoproteins in the SWISS-PROT database.* PROTEOMICS, 2001. **1**(2): p. 262-268.

6. Farriol-Mathis, N., et al., *Annotation of post-translational modifications in the Swiss-Prot knowledge base.* PROTEOMICS, 2004. **4**(6): p. 1537-1550.

7. Khoury, G.A., R.C. Baliban, and C.A. Floudas, *Proteome-wide post-translational modification statistics: frequency analysis and curation of the swiss-prot database.* Sci. Rep., 2011. **1**.

(564d) Genomic Approaches to Evolution and Adaptation of Modern Human Populations

Lawrence I. Grossman[1], Denis Pierron[2], Thierry Letellier[3], Nicolás Gutiérrez Cortés[3], Harilanto Razafindrazaka[3] and Christophe Rocher[3], (1)Center for Molecular Medicine and Genetics, Wayne State University, Detroit, MI, (2)Wayne State University, Université Victor Segalen Bordeaux 2 , Bordeaux, France, (3)Université Victor Segalen Bordeaux 2, Bordeaux, France

Our understanding about the forces that have shaped human genome evolution has accelerated during the last years based on the study of the human diversity utilizing large genomic datasets. The availability of data from 11 diverse populations has allowed us to examine the role of genetic variation in population migration, organ differentiation, and the process of evolution itself, and we present here an illustration of each. In terms of migration, previous studies have focused on mitochondrial DNA (mtDNA) polymorphisms and have shown astonishing correlations between individuals' geographic origin and mitochondrial haplogroup, revealing the migration and settlement history of human species. Based on a world sampling of 629 unrelated individuals, we have now studied the polymorphisms of the 80 genes encoding nuclear subunits of the mitochondrial respiratory (OXPHOS) complexes. We have shown that (i) populations can be distinguished based only on amino-acid replacements in nuclear encoded OXPHOS subunits, and (ii) amino-acid replacement frequencies are significantly correlated with their pathogenicity probability. These results, which are congruent with the major mtDNA haplogroups, suggest that OXPHOS complexes are different across the populations in both nuclear and in mitochondrial encoded subunits.

In terms of organ differentiation, studies about gene expression have suggested that some tissue-specific genes evolved rapidly on the human lineage. Thus, it appears that constraint on genes have varied during human evolution depending on the tissue in which they are expressed. To gain insight into the evolution of tissue determined negative selection forces, we compared the nonsynonymous SNP diversity of genes expressed in different tissues. For each SNP, we determined its frequency in 11 human populations and in each case predicted whether or not the change it produces is deleterious. We have shown that, with one exception, for all organs under study SNPs predicted to be deleterious are present at a significantly lower frequency than SNPs predicted to be tolerated. However, testis specific genes contain a notably higher proportion of deleterious SNPs compared to those of other organs, suggesting that negative (maintaining) selection on human testis specific genes has recently decreased.

In terms of evolution, the question of whether it is ongoing in modern human populations is one of continuing interest. To address it, we studied the selective strengths acting presently on the genome and show that, with one exception, the genetic diversity for all the major pathways is still constrained by negative selection in all 11 human populations studied. The exception was a relaxation of negative selection acting on olfactory receptors. Since a decreased number of functioning olfactory receptors in human compared with other primates had already been shown, this suggests that the role of olfactory receptors for survival and reproductive success has decreased during human evolution. By showing that negative selection is *still* relaxed, the present results imply that no plateau of minimal function has yet been reached in modern humans and therefore that olfactory capability might still be decreasing. This is a first clue to present human evolution.

(564e) Gene Network Inference and the Immune Response: A Novel Approach to Identifying Biology's Risk Assessment Strategies

Jason E. Shoemaker[1], Satoshi Fukuyama[1], Hiroaki Kitano[2] and Yoshihiro Kawaoka[1], (1)Japan Science and Technology Agency, Tokyo, Japan, (2)Systems Biology Institute, Tokyo, Japan

Gene network inference and the immune response: A novel method to identifying biology's risk assessment strategies

Daily, our immune systems protect us from the various pathogens to which we are exposed. When a pathogen, such as a virus, successfully invades our body, the immune system can employ various biological processes to contain the infection but these processes are also capable of invoking a great deal of tissue damage (e.g., inflammation). Thus, an effective immune response must assess the risk pathogens pose to our health and carefully manage the magnitude of its response to minimize collateral damage. A mismanaged response has been implicated to result in a more sever pathological outcome during virus infections[1–3] and is the underlying cause of allergic reactions[4]. Thus, a better understanding of the mechanisms and strategies behind how the immune system detects a pathogen and manages its response will greatly facilitate the design of immune-intervention therapies and may identify novel, host-based drug targets.

Recently, our group has applied gene network inference algorithms to dynamic, *in vivo* gene expression data collected from influenza infected mice to determine which biological processes track the state of the infection, how they communicate with other processes responsible for virus clearance and whether the communication between the various processes is dependent on the virulence of the pathogen. We applied a series of gene set enrichment tests, including our recently developed cell-type enrichment platform, CTen, to identify if a module may be regulated by a particular biological process or if the module represents conserved gene expression resulting from lymphocyte infiltration. Lastly, by analyzing the correlation between gene modules (as opposed to the traditional approach of analyzing within a module to infer gene to gene connectivity), we developed a network abstraction of the immune response to determine how various biological processes are coordinated during infection.

We found that several major aspects of the immune response, from virus detection events (such as Toll-like receptor signaling) to infection management events (e.g., intracellular death via apoptosis or targeted cell death using lymphocytes), were apparent in microarray data as modules of coexpressed genes. Using flow cytometry, we validated that CTen correctly identified gene expression that is the result of the influx and exodus of macrophages, B cells and T cells.

Focusing of the dynamics of gene expression in each module, we found that some host pathways employ ultrasensitive mechanisms (low levels of pathogen do not evoke a response, but a strong response occurs after a threshold level is reached). Given the high noise environment in which the immune system operates, we hypothesize that these mechanisms aid in effective noise rejection and play a critical role in the risk assessment apparatus.

Then, expanding our analysis to the network level, we found that the connectivity pattern between the modules was very consistent with the known behavior of the immune response, but, we also observed several modules which appear to be critical for the successful transition from the innate to the adaptive immune response. Many of these modules were related to processes which are not generally associated with the immune response while others had no known biological function. We are now performing additional experiments to determine if these apparent interactions represent causal relationships.

In sum, we present a new approach to profiling the immune response which offers several opportunities for the development of immunomodulatory therapies. We validated that immune cell infiltration can be determined from tissue array data using our CTen platform, identified control strategies employed by the immune response to gauge when an invading pathogens necessitates a response, and found several targets to be considered for targeted intervention.

REFERENCES

1. Mohamadzadeh, M., Chen, L., Olinger, G.G., Pratt, W.D. & Schmaljohn, A.L. Review Filoviruses and the Balance of Innate, Adaptive, and Inflammatory Responses. *Viral Immunology* **19**, 602-612 (2006).

2. Cilloniz, C. *et al.* Lethal dissemination of H5N1 influenza virus is associated with dysregulation of inflammation and lipoxin signaling in a mouse model of infection. *Journal of virology* **84**, 7613-24 (2010).

3. Kobasa, D. *et al.* Aberrant innate immune response in lethal infection of macaques with the 1918 influenza virus. *Nature* **445**, 319-23 (2007).

4. Boyce, J. a, Bochner, B., Finkelman, F.D. & Rothenberg, M.E. Advances in mechanisms of asthma, allergy, and immunology in 2011. *The Journal of allergy and clinical immunology* **129**, 335-41 (2012).

(564f) Combinatorial Transcriptional Regulatory Network Driving Aberrant Effects of Chronic Alcohol Consumption On Liver Regeneration

Daniel Cook[1,2], Lakshmi Kuttippurathu[3], Biswanath Patra[3], Jan Hoek[3], Babatunde A. Ogunnaike[1] and Rajanikanth Vadigepalli[2], (1)Department of Chemical and Biomolecular Engineering, University of Delaware, Newark, DE, (2)Daniel Baugh Institute for Functional Genomics and Computational Biology, Department of Pathology, Anatomy and Cell Biology, Thomas Jefferson University, Philadelphia, PA, (3)Department of Pathology, Anatomy and Cell Biology, Thomas Jefferson University, Philadelphia, PA

Chronic alcohol intake interferes with the onset and progression of liver regeneration after tissue damage; suppression of this repair mechanism is thought to contribute to the development of alcoholic liver disease. Our overall objective is to uncover and characterize the regulatory network affected by chronic alcohol consumption and how alterations in this network lead to an aberrant liver repair and regeneration response. Our strategy is to utilize a combination of experimental analysis of gene expression and transcription factor activity changes combined with computational modeling to evaluate the impact of chronic ethanol consumption on the transcriptional regulatory network. In the present study, we aimed to identify the genes and regulatory patterns associated with the early stage of liver regeneration after partial hepatectomy (PHx) to examine the effects of chronic ethanol intake on this process.

We employed the Lieber-DeCarli alcohol pair-feeding protocol to assess the extent to which the molecular processes underlying liver regeneration in alcohol-fed rats are similar to pair-fed littermate controls in which ethanol calories were replaced by carbohydrates. Our previous studies showed that the alcohol-fed group contained significant changes to transcription factor (TF) localization during the priming phase of liver regeneration (0-6 hours). Genome-wide binding targets of multiple TFs (of NFκB, C/EBPβ, C/EBPα, STAT3) were detected using either Roche NimbleGen ChIP-ChIP microarray platform (ChIP-chip) or ABI SOLiD platform (ChIP-Seq). For each TF, we annotated the localization site data with TF binding site (TFBS) genes and discretized the TFBS data by specifying if a gene was bound by the TF (1) or not bound (0). In order to minimize technical noise, each TFBS was considered to be occupied only if a majority of the biological replicates showed statistically significant binding. We integrated results from multiple TFBS data into dynamic binding activity patterns reflecting the effect of an alcohol diet on combinatorial TF binding during the priming phase of liver regeneration. We examined each pattern for enriched pathways to identify alcohol-affected biological functions.

Our results revealed dynamic changes in binding of several early response TFs at the promoters of key genes differentially co-regulated during alcohol exposure and the onset of liver regeneration. Analysis of binding patterns indicated significant differences in the genome-wide targeting of NFkB activity between chow-fed and isocaloric pair-fed rats. We identified target genes unique to the pair-fed group and unique to the chow-fed group. The former set included genes associated with functions such as positive regulation of caspase activity, protein complex assembly, and apoptosis. The target gene set unique to chow-fed rats included genes associated with functions such as signaling, positive regulation of organelle organization, and positive regulation of cell growth and size. We also identified genes with a delayed increase in NFkB binding activity in the chow-fed group, including genes related to cell adhesion, and genes with an early increase in NFkB binding activity, including genes related to the cell cycle. Given the similarity of physiological response in the two dietary groups, it is likely that these differences in NFkB binding are compensated for elsewhere in the global regulatory network driving liver regeneration.

The binding activity dynamics of the TFs of NFκB, C/EBPβ, C/EBPα, and STAT3 at target genes were altered by chronic alcohol intake. Further analysis of the dynamic patterns led to novel insights about the differences in the transcriptional network subtending liver regeneration between alcohol-fed rats and pair-fed controls. The dominant patterns were either novel in the alcohol group or unique to the control group. Functional association of genes in these key dynamic response patterns revealed pathways such as cellular homeostasis, cell cycle regulation, RNA metabolism, and translation, to be affected by novel and missing activity in the alcohol group. Together, these results indicate that the effects of alcohol on liver regeneration may be mediated by dysregulation of cell cycle related processes and activation of alternative pathways.

Promoter analysis by PAINT revealed that several target genes may be driven by regulatory modules containing co-located binding sites for multiple TFs. We validated a subset of 20 NFκB targets and 5 key target promoters for the other TFs using ChIP-qPCR. Our results reveal that: (1) increased baseline activation of NFκB, C/EBPβ, C/EBPα, STAT3, c-Fos, and c-Jun in regenerating livers is correlated with the promoter binding for a subset of the target genes; (2) similarly, alcohol-fed rats showed a further increase in NFκB binding activity in response to PHx at 6 hours post-PHx occurring at only a subset of target promoters as opposed to pair-fed controls, which showed a consistent increase at all promoters tested during early liver regeneration; and (3) C/EBPβ, C/EBPα, STAT3, c-Fos, and c-Jun binding activity were down regulated at a subset of target promoters in alcohol-fed rats during early liver regeneration.

We conclude that chronic alcohol intake significantly affects the system-wide transcriptional regulatory network, including TF binding to the key genes active during the early stage of liver regeneration. These alterations may be critical in disrupting the normal regulatory network dynamics driving effective liver regeneration.

AUTHOR INDEX

AUTHOR INDEX

AUTHOR INDEX